Conceptual Change and the Philosophy of Science

W0018679

In this book, David Stump traces alternative conceptions of the a priori in the philosophy of science and defends a unique position in the current debates over conceptual change and the constitutive elements in science. Stump emphasizes the unique epistemological status of the constitutive elements of scientific theories, constitutive elements being the necessary preconditions that must be assumed in order to conduct a particular scientific inquiry. These constitutive elements, such as logic, mathematics, and even some fundamental laws of nature, were once taken to be a priori knowledge but can change, thus leading to a dynamic or relative a priori. Stump critically examines developments in thinking about constitutive elements in science as a priori knowledge, from Kant's fixed and absolute a priori to Quine's holistic empiricism. By examining the relationship between conceptual change and the epistemological status of constitutive elements in science, Stump puts forward an argument that scientific revolutions can be explained and relativism can be avoided without resorting to universals or absolutes.

David J. Stump is a philosopher of science at the University of San Francisco, San Francisco, California. He is coeditor, with Peter Galison, of *The Disunity of Science* and is author of numerous journal articles on the history and philosophy of science.

Routledge Studies in the Philosophy of Science

Conceptual Change and the Philosophy of Science

Alternative Interpretations of the A Priori

David J. Stump

Routledge
Taylor & Francis Group

LONDON AND NEW YORK

First published 2015
by Routledge

2 Park Square, Milton Park, Abingdon, Oxfordshire OX14 4RN
711 Third Avenue, New York, NY 10017

Routledge is an imprint of the Taylor & Francis Group, an informa business

First issued in paperback 2018

Library of Congress Cataloging-in-Publication Data

Stump, David J.
    Conceptual change and the philosophy of science : alternative
interpretations of the a priori / by David J. Stump.
        pages cm. — (Routledge studies in the philosophy of science ; 16)
    Includes bibliographical references and index.
    1. Science—Philosophy.   I. Title.
    Q175.S934 2015
    501—dc23
    2015002885

ISBN: 978-1-138-89013-8 (hbk)
ISBN: 978-1-138-34669-7 (pbk)

Typeset in Sabon
by Apex CoVantage, LLC

To Andrea, for everything.

Contents

Figures

Preface

Conceptual change during scientific revolutions is a major topic in the philosophy of science, especially in the wake of Kuhn's *Structure of Scientific Revolutions*. It has been suggested by many philosophers that when the constitutive elements of scientific theories change, there is a conceptual revolution. By constitutive elements, I mean necessary preconditions for conducting a particular scientific inquiry. For example, much of modern science is formulated in mathematical terms, so the mathematics is therefore constitutive of the scientific theory. The discovery of non-Euclidean geometries in the nineteenth century had a profound impact on philosophy and paved the way for conceptual revolutions in physics in the twentieth century.

Constitutive elements were often taken to be a priori. Indeed, one way to organize much of twentieth-century philosophy of science is to read it as a series of debates over what had been considered a priori knowledge. Although synthetic a priori knowledge was officially rejected by the Vienna Circle and their followers, parts of knowledge that had been considered a priori by Kant, such as geometry, space and time, causality, and the basic principles of physics, were widely discussed throughout the twentieth century, and statements about them have often been given a special role, whether as conventions or as the hard core of scientific theories. One would think that Quine's critique of the analytic/synthetic distinction would have put the final nail in the coffin of a priori knowledge and that a holism in which all scientific statements are justified empirically has replaced the notion of any special status for what was formerly considered to be a priori, but in fact, replacement theories of the a priori were developed throughout the twentieth century by Hans Reichenbach, C. I. Lewis, Ernst Cassirer, Arthur Pap, and, most recently, Michael Friedman. Even though Kant thought of a priori knowledge as fixed and absolutely certain, these theories see the (former) a priori as changing during scientific revolutions.

While it is possible to defend Carnap or even Kant from Quinean holistic empiricism, I take a different approach, arguing that the former a priori should be treated not only as empirical in a very abstract sense but also as constitutive, by which I mean that some elements of physical theory have a unique epistemological status in that in order to begin empirical inquiry one must adopt these elements of science, which therefore function as a priori parts of our physical theory, chosen for conceptual or pragmatic reasons prior to any empirical testing. Although I am willing to concede to Quine that these constitutive elements of science are ultimately empirical, the picture of knowledge developed here is very different from that developed in Quinean holism in that categories of knowledge can be differentiated. While he admits that some elements of empirical theory are much less likely to be revised than others, differentiating between what he calls the hard core and the periphery, Quine underestimates the asymmetric relation between the constitutive elements and the rest of science. It is not simply that the periphery is more likely to be revised than the hard core but, rather, that the statements of the periphery cannot even be stated, let alone tested, without the constitutive elements functioning as a necessary precondition.

By tracing the development of alternative conceptions of the a priori in the philosophy of science, I defend a unique position in the current debates over the constitutive elements in science, following and expanding on the position that I set out in "Defending Conventions as Functionally A Priori Knowledge" (*Philosophy of Science* [2003] 20: 1149–1160). My position has evolved considerably and drops the use of 'a priori' altogether, despite the fact that it has been traditional in this literature, to suggest that we consider the nature of theories of the constitutive elements in science. These range from Kant's view that the constitutive elements are fixed and universal, to views like Pap's, in which what is constitutive changes from one context to another. I argue that theories of the constitutive elements in science can be formulated without falling into relativism, idealism, or social construction.

I first presented parts of the book as talks, and I thank the hosts for the opportunity and members of the audience for helpful discussion. The opening section of Chapter 3 was presented at the Fourteenth Congress for the Logic, Methodology and Philosophy of Science, Nancy, France, July 25, 2011, while an early version of Chapter 4 was given as a talk at the Eighth International HOPOS Conference, Budapest, Hungary, June 25, 2010. Versions of the material on C.I. Lewis in Chapter 5 was presented at the Tenth International HOPOS Conference, Ghent, Belgium, July 4, 2014, and at the Pragmatism and Science Conference, Amherst, New York, June 20, 2009. Parts of the material on Pap in Chapter 5 were presented at the University of Vienna, May 16, 2013, the Ninth International HOPOS Conference, Halifax, Nova Scotia, Canada, June 22, 2012, and the Seventh International HOPOS Conference, Vancouver, British Columbia, Canada, June 2008. Part of Chapter 7 was read at the University of Vienna, May 14, 2013. Chapter 8 was presented to audiences at San Francisco State University on February 28, 2013, and at the University of Lorraine, September 22, 2014.

I benefited greatly from a stay as a visiting fellow at the Center for Philosophy of Science, University of Pittsburgh, during the fall of 2012 and from having the chance to present and discuss my work on the book. I thank especially the director, John Norton, for his hospitality, organizational skills, and the supportive atmosphere he has created at the center, as well as the other fellows and postdocs, Marta Bertolaso, Maria Kronfeldner, Dennis Lehmkuhl, Alexander Reutlinger, Collin Rice, Kyle Stanford, and Şerife Tekin, for their comradery and feedback. I also thank Nicholas Rescher for his interest and hospitality, Robert Batterman, and James Lennox.

I benefited from a prolonged research stay at the Archives Henri Poincaré at the University of Lorraine in the fall of 2014 and would like to thank Gerhard Heinzmann for the invitation and for his collaboration and friendship. I also benefited from interaction with Scott Walter, Philippe Nabonnand, Pierre Edouard Bour, and Kate Hodesdon.

Many others have been helpful to me in various ways over the years, through discussions at conferences and presentations and through correspondence. I would like to thank Hasok Chang, Gerald Doppelt, Massimo Ferrari, Arthur Fine, Menachem Fisch, Janet Folina, Ian Hacking, Jeremy Heis, Björn Henning, Don Howard, Arezoo Islami, Alex Klein, Martin Kusch, Tom Oberdan, Flavia Padovani, Paolo Parrini, Maria de la Paz, Chris Pincock, George Reisch, Alan Richardson, Joseph Rouse, Tom Ryckman, Eric Schliesser, Warren Schmaus, Sanford Shieh, Ed Slowik, Friedrich Stadler, Jonathon Tsou, Thomas Uebel, Bas van Fraassen, and Susan Vineberg. I thank Milena Ivanova especially for reading and helpfully commenting on the complete manuscript of the book. I thank my colleagues at the University of San Francisco for their support and encouragement, especially Manuel Vargas and Tony Fels, who commented on early versions of parts of the book. I received financial support for travel and research from the Faculty Development Fund at the University of San Francisco. The three anonymous reviewers for Routledge provided comments that improved the book considerably. Finally, I want to thank my wife, Andrea Smith, who has supported me in every possible way for many years and kept me going and who has also served as the best editor imaginable, improving the writing through many drafts. It is to her that I dedicate the book with much love and gratitude.

Acknowledgments

Figure 1.1 is used by permission of Oxford University Press. Source: *Philosophy of Science Matters: The Philosophy of Peter Achinstein*, edited by G. Morgan (Oxford: Oxford University Press, 2011), 170, figure 13.2. Figure 2.3 is used by permission of the American Mathematical Society. Source: "Figure 1.1: The Pseudosphere" in John Stillwell, *Sources of Hyperbolic Geometry* (Providence, RI: American Mathematical Society, 1996), 1.

Chapter 2 draws on material from "Reconstructing the Unity of Mathematics circa 1900," *Perspectives on Science* (1997) **5** (no. 3): 383–417, and "Poincaré's Curious Role in the Formalization of Mathematics" in *Henri Poincaré: Science and Philosophy, International Congress*, edited by J.L. Greffe, G. Heinzmann and K. Lorenz (Paris and Berlin: Albert Blanchard and Akademie Verlag, 1996), 481–492, both in substantially revised form. The first appears by permission of MIT Press, the second by permission of Walter de Gruyter GmbH. Chapter 3 draws on material from "Henri Poincaré's Philosophy of Science" *Studies in History and Philosophy of Science* (1989) **20**: 335–363, and it is used here by permission of Pergamon Press/Elsevier. The third section of Chapter 5 is largely reprinted from "Arthur Pap's Theory of the Functional A Priori" *HOPOS The journal of the International Society for the History of the Philosophy of Science* **1**, no. 2 (2011): 273–290, and appears here by permission of the University of Chicago Press. Some material in Chapter 6 is from "A Reconsideration of Newton's Laws" in *What Place for the A Priori?* edited by M. Veber and M. Shaffer (Chicago: Open Court, 2011), 177–192, and appears here in revised form by permission of Open Court Press. Part of the section on Kuhn in Chapter 7 first appears as the "Paradigm" entry in the *New Dictionary of the History of Ideas*, Charles Scribners Sons, © 2005 Gale, a part of Cengage Learning, Inc. Reproduced by permission. www.cengage.com/permissions. Chapters 7 and 9 each draw on small parts of my article "Fallibilism, Naturalism and the Traditional Requirements for Knowledge," *Studies in History and Philosophy of Science* **22** (1991): 451–469, by permission of Pergamon Press/Elsevier. A small section of Chapter 8 draws on material published in "Review of Michael Resnik, *Mathematics as a Science of Patterns*," *History and Philosophy of Logic* **19** (1998): 176–177, by permission of Taylor and Francis.

In a few cases, I have quoted more from published sources than may be called fair use. I have explicit permission to quote excerpts from Norwood Russell Hanson *Patterns of Discovery: An Inquiry into the Conceptual Foundations of Science* (Cambridge: Cambridge University Press, 1958) by permission of Cambridge University Press; from Jeremy Heis, "Realism, Functions, and the A Priori: Ernst Cassirer's Philosophy of Science," *Studies in History and Philosophy of Science Part A* **48**, no. 0 (2014): 10–19, by permission of Pergamon Press/Elsevier; and from Arthur Pap, *The A Priori in Physical Theory* (New York: King's Crown Press, 1946), by permission of Pauline C. Pap (Mrs. Arthur Pap).

1 Introduction
Theories of the Constitutive Elements in Science

[T]he possible orderings of experience are limitless; we force upon the subject matter of physics the ordering we choose.

—Hanson (1958, 97)

Suppose a researcher in a biology lab opens a new box of Petri dishes containing agar prepared for growing cultures. The researcher finds that one of the dishes contains a growth and throws it away. What are we to think about the growth in the brand new Petri dish? We might just assume, like the researcher, that the dish was contaminated somehow in the manufacturing process, allowing for the entry of microbes and their growth in a medium that was supposed to be sterile. However, we might formulate an alternative explanation of what happened, saying instead that the growth is the result of spontaneous generation in a sterile medium. The researcher has, by implication, rejected the second explanation without thinking about it at all, when throwing the Petri dish in the garbage, thereby foregoing a chance for a Nobel Prize. How can the hypothesis that the growth in the Petri dish is the result of spontaneous generation be dismissed so easily and conclusively? The answer lies in the fact that we are no longer collecting evidence for and against the theory of spontaneous generation, given that the dispute over spontaneous generation, once a live issue, was settled long ago.

The nonexistence of spontaneous generation is already settled science, which will remain settled until there is some good reason to open the question again. It was settled when Pasteur and others showed that all putative cases of spontaneous generation could be explained by the existence of microbes in the environment, so we assume that there must have been a contamination, that is, the accidental inclusion of microbes in the new Petri dish. Is it possible to raise the issue again and take the existence of spontaneous generation seriously? Theoretically speaking, reopening this or any other question in science is certainly possible, given that all results in science are fallible. There is nothing in the logic of the situation to rule out spontaneous generation as an explanation of the phenomenon. Furthermore, explanation of the origin of life in scientific terms must include some instance of spontaneous change from chemicals to life, so the idea of spontaneous generation in itself is not scientifically disreputable, it is just assumed to have taken place over millions of years during the remote past, not in contemporary laboratory Petri dishes. Practically speaking, however, reopening the spontaneous generation debate in this case is not possible. First, it would take much more than one Petri dish, or even a box of Petri dishes, to reopen the dispute. The growth of microbes in sterile Petri dishes would have to be repeatable, and alternative explanations such as contamination would have to be ruled out. Second, a hundred or more years of science would need to be reevaluated.

There is more to this case than a simple dismissal of an alternative hypothesis, however. We may say that the presence of growth in the agar in the Petri dish has become a criterion for the contamination of the dish, or as I say later, the nonexistence of spontaneous generation is now partially constitutive of biology. The researcher checks to see if the Petri dish has been contaminated by checking for growths and assumes that the Petri dish is contaminated if there are. The nonexistence of spontaneous generation has thus become a principle that is taken for granted and used to set up a criterion for the sterility of the Petri dish. When we see growth on a new Petri dish, we throw it out. Were someone to ask, we would argue that contamination of the Petri dish is much more likely than spontaneous generation, although either hypothesis would explain the growth of microbes on the dish.

There are at least two ways of looking at the setting up of criteria in this manner. The philosopher Arthur Pap discusses this in terms of meaning, so that he would want to say that the definition of an uncontaminated Petri dish includes the idea that there are no microbes in the dish before they are placed there in the course of research. We could, however, leave out any discussion of the meaning of terms and simply say that the presence of microbes in a new Petri dish has become a criterion of the contamination of the dish. The second option focusses on the practice of science, rather than semantics, which provides a picture of science that is more historically accurate.

We can take settled science for granted and treat the propositions of settled science as if they were certain, although we know that nothing is certain in science. Settled science is one kind of example of something that functions as what formerly had been taken to be a priori knowledge; that is, in the context of current research, it is taken for to be true without any examination and prior to conducting any further research. The Petri dish has to be sterile to conduct whatever experiment is being done, so it is a precondition of any further inquiry. With this example, I hope to introduce the debate over what I call constitutive elements in science. These function as a priori elements in our theory, but in a strong sense, there is no a priori. I accept the modern empiricist view that all knowledge of the world ultimately comes from sensation. There is a lot packed into the word *ultimately*, however, and my point in raising the issue is that it is far more important to recognize the role of the assumptions that are used in science, any science, than it is to simply say that everything is empirical. Saying no more than all scientific knowledge is empirical conceals what is very important about the epistemological status of the fundamental parts of our scientific theories. The Petri dish is but one kind of example of what I will be calling the constitutive elements in science.

In looking again at the functional and the related pragmatic conception of the a priori, I am tracing the history of an idea that has gone into disuse. It was never a popular idea but was always intellectually respectable. It was discussed surprisingly little, given that it is an important variation of the philosophy of science, one that shares its roots with American Pragmatism, rather than the Logical Empiricism that was imported into the United States by the Vienna Circle and its allies. It is unfortunate that the discussion has so often been linked to a priori knowledge because this leads to some confusion. The pragmatic or dynamic a priori is so unlike the traditional idea of a priori knowledge that it really deserves a new name, the 'constitutive elements of science.' As with the Kantian a priori, functionally (or pragmatic or dynamic) a priori knowledge can be said to be constitutive of a science; unlike Kant, functionally a priori knowledge is relative and contingent and has changed historically. The distinction between the two meanings of a priori knowledge in Kant—the necessary and the constitutive—and the focus on the relativized or dynamic constitutive elements goes back to Hans Reichenbach ([1920] 1965) and has recently been elaborated and defended by Michael Friedman (2001, 2002, 2003a, 2003b, 2004, 2005a, 2005b, 2006, 2008a, 2008b, 2009, 2010a, 2010b, 2011, 2012). In one period Hilary Putnam also developed the notion of a relativized a priori in order to address the problem of scientific change (Tsou 2010). Reichenbach distinguishes the constitutive and necessary, eternal elements of the a priori, thus dividing the term *a priori* into two elements. I would like to suggest instead that we think of the term *constitutive* as primary and as consisting of two kinds of theories, those such as Kant's that take the constitutive elements in science to be a priori in a traditional sense and those that take the constitutive elements of science to be dynamically changing.

It is certainly true that the constitutive elements in science frequently turn out to be things that were once considered to be a priori knowledge in a full-bodied sense, but the adjective in front of the word *a priori* in these theories totally changes the meaning of the term. The dynamic, pragmatic, functional, and relativized a priori are not actually theories of a priori knowledge but, rather, are theories of the constitutive elements in science. They give a new epistemological status to things that had formerly been called a priori. Nevertheless, there are three ways that the theories of the constitutive elements in science are connected to the traditional idea of a priori knowledge. First, many of the examples come from things that were considered a priori, especially by Kant. Our conception of space and time, our knowledge of mathematics, and a few fundamental principles of physical theory would fall into this category. Second, those advocating theories of the constitutive elements in science have emphasized that some of knowledge must be in place prior to our being able to conduct scientific research in a particular way. These are necessary preconditions for the possibility of conducting a scientific inquiry, to put it in Kantian language. For example, the mathematics has to be known before the empirical inquiry can begin, given that theories are stated in mathematical language and that problems are solved using the tools of mathematics. Third, many aspects of scientific knowledge are taken for granted as established, and furthermore, some of these aspects can be taken as a criterion for further inquiry. These principles are not only 'hardened' so that no one would seriously doubt them but also play a special role in categorizing phenomena.

As I am using the term, constitutive elements in science play a role that is somewhat different from the usage in Kant. My focus is on what makes a particular scientific practice possible, and I argue that there are necessary preconditions that I will call constitutive. I am not suggesting that something a priori in the mind of the subject constitutes an object. That position, which skids into idealism, was explicitly rejected by C. I. Lewis when distinguishing his theory of the pragmatic a priori from that of his teacher Royce (Dayton 2006). The issue of whether objects are constituted comes up in Schlick's discussion of Reichenbach's early view (Oberdan 2009), and we see it again in Kuhn's idea of world change (Kuhn 1962).[1]

As I claimed, these theories of the constitutive element form an alternative tradition in the philosophy of science. In tracing the history of the philosophy of science in a narrow sense, I look to America in the late 1930s for the origins of the discipline, and at the 1950s for its full establishment. The philosophy of science was virtually invented as a subspecialization in philosophy by the Logical Empiricists. I recognize that philosophers have always had things to say about science, but the point here is tracing the invention of philosophy of science as a discipline, as a specialization in philosophy. I do not think that the philosophy of science in this sense was created until the Philosophy of Science Association was established. I am talking about more than the organization when I talk about the philosophy of science as a discipline, but I mean something narrower than general philosophical reflections on science. By a philosopher of science, I mean someone whose main philosophical output relates to philosophical questions of science. The story told here is an alternative to mainstream philosophy of science, one that was influenced by pragmatism. Arthur Pap, the inventor of the term *functional a priori* plays an important role in my history. What is interesting about Pap is that while he was a fellow traveler in many senses to the Logical Empiricists, his philosophy of science is strongly influenced by pragmatism. There is a direct line of descent from Pap to C. I. Lewis, Dewey, and from Dewey to Peirce, and another neo-Kantian set of influences, mostly from Cassirer. Victor Lenzen inspired Pap to apply Poincaré's conventionalism to issues in the philosophy of science, so, Poincaré is a natural place to start for the development of these ideas.

THE CONTEXT OF INQUIRY

All knowledge is based on some assumed starting point. In science, there are principles and theories that are taken for granted before empirical inquiry can begin. Although these theories and principles may have been confirmed empirically, some fundamental principles or laws and all of the mathematics on which science depends have a more problematic basis. We find that these principles are very difficult to conceive of as being empirically grounded. Some of the principles and all of the mathematics look like a priori knowledge—independent of experience and a necessary precondition to further inquiry.

In the twentieth century, philosophers of science moved firmly away from the rationalist idea that there is a priori knowledge that is known by some kind of intuition. Nevertheless, it is striking how central topics in early-twentieth-century philosophy of science match up with things that Kant considered to be synthetic a priori knowledge: space, time, and causality (and, by extension, explanation). The mainstream view was a strict empiricism in which all a priori knowledge was deemed to be analytic, but there is a different strand of thinking in twentieth-century philosophy of science that acknowledged the special role that what had formerly been called a priori knowledge plays in science. Poincaré's geometric conventionalism, Cassirer's neo-Kantianism, Reichenbach's constitutive a priori, Lewis's pragmatic conception of the a priori, Pap's functional a priori, and Kuhn's paradigms all fall into this camp. This book traces the history of these alternative positions and argues for one coherent understanding of an alternative to mainstream logical empiricism.

These problematic elements of scientific knowledge were exactly the ones that changed during scientific revolutions. For example, Kant's theory of space and time was overthrown in Einstein's theory of relativity. Therefore, looking at the constitutive elements in science gives us a way of understanding deep conceptual change in science, which again seems to be neither simply empirical nor unproblematically rationalistic. Scientific revolutions occur when there is a change in constitutive elements in science, that is, a change in what had been taken to be a priori knowledge.

It is true, of course, that the genesis of the philosophical concern over spacetime arose because of Einstein's tremendous influence on the Logical Empiricists and their further influence in defining the philosophy of science in the twentieth century. Schlick, Carnap, and Reichenbach each wrote philosophical responses to Einstein's theory of relativity. The Logical Empiricists explored the idea that all of formerly a priori knowledge is analytic, but this position was shown to be problematic and many consider it to have been refuted by Quine, who rejected truth by convention and the very distinction between analytic and synthetic statements, adopting a holism in which all of science is empirical.

Here I set out and defend a role for constitutive elements in science, a pragmatic view that there are principles and theories that are necessary preconditions for the possibility of a science, but that stays closer to naturalism than the neo-Kantian position advocated by Friedman, who also defends a role for a dynamic, constitutive a priori in opposition to Quine's holism. Mathematics and some scientific principles play an essential constitutive role in science. They function as a priori knowledge and are indeed necessary preconditions to scientific knowledge. While these fundamental scientific beliefs appear to be a priori rather than empirical, I argue that they can in fact be considered empirical in a very abstract sense. What is functioning as a priori knowledge is not actually a priori in the traditional sense, but there is nevertheless a very important reason why this terminology was used by so many authors. It strongly underscores the special role that the constitutive elements play in science. I think that we can now give up the label a priori and leave in place the idea of constitutive elements in science as the central focus. These constitutive ideas play a special role in scientific revolutions, for they are the basis of conceptual change.

THE CONSTITUTION OF SCIENTIFIC KNOWLEDGE

By constitutive, I mean the idea that some elements of physical theory have a unique epistemological status—one must adopt these elements in order to begin inquiry. They therefore function as an a priori part of our physical theory, no matter whether their origin is ultimately empirical or is a matter of definition. There are at least three types of constitutive elements in science: those that are strictly mathematical, those that are the basic principles or laws of an empirical theory, and former empirical statements that have become fixed. As an example of the first type, the calculus is necessary to even formulate the contemporary version of Newtonian theory, because the idea of instantaneous velocity cannot be formulated without the calculus (Friedman 2002, 178).[2] There are, of course, countless examples in which scientific theories are formulated in a mathematical language and where we could rightly say that we could not even begin to do the science without the mathematics—vector algebra is required for even high school physics, probability theory is required for population genetics, and Riemann's results in geometry are required for the General Theory of Relativity. For the second type of constitutive element in science, examples of physical laws that are required before a science would be possible, there are Newton's laws of motion themselves, the inverse square law, and the principle of natural selection. We cannot even begin to study Celestial Mechanics without Newton's laws of motion and the law of gravity, for example.

An example of what has been called 'Kuhn loss' will be helpful in seeing how the constitutive elements can change and the empirical, while not unaffected, does not simply change with them. Consider Newton's principle of inertia: Before Newton (and Galileo) the question was what keeps an object in motion, with so-called violent motion requiring force. When force is removed, the object keeps going for a while but then stops, a phenomenon that was explained with the theory of impetus. The crucial point here is that Galileo and Newton changed the question. Rather than ask what keeps an object in motion, the question for them is what changes a body's state of motion, since continued motion in a straight line is taken to be natural and does not need to be explained (except to the extent that Newton's first law can be called an explanation). Whereas for the impetus theory, continued motion in a straight line required an explanation and gets one, namely, either that force must be continually applied, like in the case of a horse drawing a carriage, or that impetus is imparted to an object and continues its motion for a while, until it runs out, as in the case of a cannonball. The impetus theory had an answer to a question that no longer exists; this is what is meant by Kuhn loss, and it is an important example of conceptual change in science.[3]

Another example of a specific presupposition in a physical theory comes from John Norton and John Manchak (Manchak 2009; Norton 2011). In the General Theory of Relativity, spacetime is extendable in various ways but we cannot tell, from what we have observed so far, what the rest of spacetime is like. So, we need to make an assumption that spacetime is well behaved, that it will not turn out that we have what Manchak calls a nemesis spacetime. Norton provides a simple example with the following two scenarios. On the left we have regular Minkowski spacetime and on the right half spacetime, with everything from t = 0 being cut off. That is, the universe just ends when we get to t = 0.

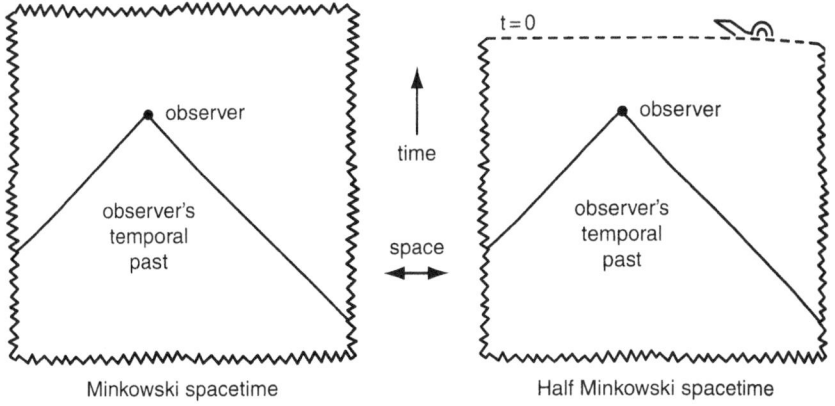

Figure 1.1 Minkowski Spacetime and Half Spacetime

Source: Philosophy of Science Matters: The Philosophy of Peter Achinstein, edited by G. Morgan (Oxford: Oxford University Press, 2011), 170, figure 13. Reprinted by permission of Oxford University Press.

There is no way to rule out the ending of the universe other than the positing of a rule that the universe will not be extended in such a fashion. There are less extreme cases to rule out as well, such as a hole in future spacetime.

It may be helpful to point out that, in Norton's view, inductive inference is theory specific. In a book in preparation, he calls this a material theory of induction.[4] Rather than a general principle of the regularity of nature, we have specific presuppositions that are necessary for particular theories. It is these specific inductive presuppositions that are constitutive of theory and of inquiry, I want to say, in the sense that such presuppositions are necessary before we can proceed. The crucial thing in Norton and Manchak's examples is that while working within the General Theory of Relativity, we make a presupposition that spacetime is well behaved, even though our evidence so far is compatible with lots of different alternatives. We cannot test our presupposition (although it is testable in principle if we wait long enough), but we cannot proceed without the presupposition. The role of what I would call physical presuppositions is likewise not so easily modeled in Quine's web of belief. These are certainly not well confirmed—indeed, sometimes they are not confirmed at all. They also nonempirical in the sense that our theories tell us that we have no way to test them, by their very nature, they are outside of the reach of empirical theory. Of course, we can say that they are confirmed to the extent that we have so far seemed to have a well-behaved spacetime. There are also fundamental laws of nature that are constitutive in science. For example, Michael Stöltzner (2009) has made a compelling case for the principle of least action.

We see therefore that there are at least three kinds of cases of the constitutive elements in science: The fixing of formerly empirical statements which are turned into criteria as in the case of spontaneous generation, the necessary preconditions in the sense of tools needed to start inquiry (e.g., mathematics) and presuppositions about what the physical world is like, such as the law of inertia or the extendibility of spacetime (also needed to begin inquiry). In some cases these might have once been considered a priori, but I suggest that they are now better seen as (logically) prior conditions on inquiry, that is, the constitutive elements of science. Friedman complains, with a certain amount of justice, that the fixing of formerly empirical statements is the same as Quine's hard core (Friedman 2001, 88n22), but there are several differences that are worth noting. First of all, it is not necessarily the case that these statements are better confirmed. They may be taken as fixed for practical reasons. Furthermore, no matter what the origin, their fixing and use as a criterion is not quite the same as what Quine describes as their position in the web. Quine makes it sound like if we get a disconfirmation, we question the beliefs at the periphery before those in the center. In the case of the fixed principles, we need a conscious revolution before we change these, not just an adjustment of our beliefs. Still, I admit that a Quinean can handle cases like this more easily than the other two.

The constitutive function gives principles of physics and parts of mathematics their distinctive epistemological status, one that is missed by strict empiricism and that is masked by Quinean holism. While mathematics has traditionally been considered to be a priori knowledge by most philosophers, some laws of nature, and Newton's in particular, have been much more controversial, sometimes being considered a priori and sometimes being considered to be empirical. They are the perfect example of statements that function as a priori knowledge in the sense that they are constitutive of a natural science, while at the same time appearing to be straightforward descriptions of physical things and properties. While rationalists may have the upper hand in arguing that our knowledge of mathematics must in some sense be a priori, empiricists will certainly feel that they want to dig in their heels and insist that all of our knowledge of physical things and processes is gained through the senses, however indirectly. Rationalists will have as hard a time justifying the claim that some of our knowledge of the physical world is a priori as empiricists have justifying the claim that mathematics is a posteriori. The fundamentally important point, however, is that the laws of nature are not straightforwardly empirical. Indeed, they can easily appear to function as definitions.

MY POSITION IN THE DEBATE

I set out a novel position in the current discussion of Quinean holism and the dynamic a priori within philosophy of science. When the issue is approached pragmatically, I argue, Quine can be granted his general points that all scientific knowledge is empirical and that all of our scientific beliefs are revisable but not his view that our scientific beliefs are distinguished by their degree of entrenchment. Some of our scientific beliefs play an essential constitutive role, in that they are necessary preconditions to scientific knowledge. Quine is wrong to ignore the constitutive element of the fundamental principles of science.

My view is close to Carnap's in that the background commitments of a scientific theory determine what is a priori and what is empirical. However, Carnap makes a mistake in considering the background commitments as a language, when they are actually more like scientific theories. He sees a priori truths as analytic, leaving himself vulnerable to Schlick's critique of conventions as trivial, as well as the famous critiques by Quine. The process of determining what is a priori and what is empirical is embedded within a physical, empirical theory. It is necessary to develop a considerable amount of theory before empirical tests are possible at all, and even then, it may seem that empirical tests have very little to do with fundamental principles. Nevertheless, theories of spacetime are ultimately empirical; that is, our change from Newtonian theory to the General Theory Relativity has an empirical basis, although a rather indirect one.

Recent literature on spacetime shows that a constitutive element remains in spacetime theories because, with the proper background theory, Poincaré's conventionalism would have been possible. It is a merely contingent fact (and rather surprising too) that uniform acceleration is detectable and unrelativisable, while uniform velocity is not detectable and is relativisable. The most important point for this discussion, however, is that the epistemological status of the metric of space or spacetime changed as we changed background theories. In this case, Newtonian theory and the General Theory of Relativity each set the epistemological status of space (or spacetime). In Newtonian theory, especially as elaborated by Kant, we know a priori that space is Euclidean, but in the General Theory of Relativity, the metric is determined empirically. Background theory and context determine what is a priori and what is empirical, that is, which parts of the theory function as a priori knowledge. This "hard core," to use Quine's term, is still empirical, but only in a very indirect and abstract sense. Thus, the Newtonian theories of space and of motion are empirical theories, but the metric of space is Euclidean and functions as a priori knowledge within Newtonian theory and is therefore a constitutive element.

Some may argue that principles such as Newton's laws are just simply empirical and therefore unproblematic, even if they are constitutive. I would like to turn this argument on its head and point out they many constitutive elements of science are empirical and that this does not diminish their role as constitutive principles. Even if they start out as empirical, they function as a priori knowledge within the context of a mature scientific theory. This phenomenon has long been noted as the hardening of a principle into a definition, and Pap (1946) makes it the centerpiece of his discussion. Rather than say that an empirical law has been elevated to the status of a definition, however, I would say that a law has in this case come to function as a priori knowledge or has become a constitutive element in science.

The most interesting aspect of the constitutive elements of a science is that they can change when a physical theory changes, for a combination of empirical and conceptual reasons. In Newton's theory, space is three-dimensional and Euclidean. In the Special Theory of Relativity, spacetime is four dimensional but still Euclidean. In the General Theory of Relativity, spacetime is a four-dimensional Riemannian manifold with variable curvature. A theory can be grounded empirically but still contain functionally a priori elements than cannot be directly tested and without which the theory could not be stated. The constitutive elements are not determined a priori, or conventionally, but are rather determined empirically as embedded elements in a physical theory. For example, Newton's laws are necessary preconditions of Newton's theory of motion, thus seeming to be a priori, but at the same time, they were eventually overthrown by the Einstein's theory of relativity. Einstein's fundamental laws and the applications of mathematics that he used are justified, in part, by the empirical success of the General Theory of Relativity, showing that even the constitutive elements of a physical theory can be influenced indirectly by empirical results.

QUINE'S HOLISM

Quineans would say that what was once called a priori is simply the hard core of our empirical theories, that is, the most well-established parts of our physical theory. However, the constitutive principles are not necessarily more entrenched and frequently they have never been *directly* tested. As Friedman notes, the difference between the core and the periphery is not a difference in the degree of justification, as Quine's model suggests (Friedman 2001, 46). Indeed, the notion of degree of justification will not at all do justice to the difference between the constitutive principles of physical science and the empirical ones. There is a marked difference between the theories of constitutive elements in science that I have been defending and Quine's view, mainly because the reason that we say that something is constitutive or functionally a priori has nothing to do with entrenchment. Quine would have us believe that the core consists of statements that have been around for a long time and that are accepted by everyone, but neither of these two conditions is necessarily met by the constitutive elements in science. For example, Newton proposed the calculus at essentially the same time as his physics, but calculus is clearly a necessary precondition to working in Newtonian physics, so it counts as constitutive in that context. The calculus was not entrenched in any sense at that time, because it was brand new and it was controversial. The calculus did not become uncontroversial until after the work of Cauchy, Weierstrass, and others put it on a sound footing.

Quineans might try to fit necessary preconditions into his web of belief as well and thus account for the constitutive elements in science. The issue is building enough nuance into the Quinean web to capture the different roles that the former a priori is playing in science. Some might argue that Quine has the resources to do this, especially considering his late works (Quine 1992, 1995). For example, in discussing his maxim of minimal mutilation, Quine notes that mathematics will inevitably be spared when we try to revise our beliefs in the light of negative evidence, because "mathematics infiltrates all branches of our system of the world" (1992, 15). While that may be true of mathematics in general, there are certainly specialized areas of mathematics that affect very little of our system of the world, especially at a given point in history. Newton's calculus certainly did not infiltrate everything when it was introduced, and even if it is widely used today, it can hardly be said to be known by more than a small fraction of humanity, so even Quine's later discussions are problematic.

Furthermore, Quineans seem to assume that the possibility of alternative theories is more than a *mere* possibility. We know that all of empirical science is fallible, meaning that the most well established of our physical theories could be wrong and that alternative physical theories could be true. Thus, for example, the General Theory of Relativity could be false and a fully relational theory of space could be true, but saying that this is possible is a long way from actually working out such a theory. Even though possibilities of change always remain open, and science does change based on conceptual and empirical results, it is also remarkably stable. Observing how science works and analyzing the constitutive elements of science is a more interesting and informative project than pointing out that theories are, in a very unrealistic sense, salvageable no matter what the evidence says. Holism can degenerate into mere skepticism.[5]

I am opposing Quine not for his thoroughgoing empiricism, not for his fallibilism, nor his naturalism, nor even for his holism per se, but rather for the fact that his holism is far too simplistic. I agree with Friedman's argument that holism can cover up important aspects of scientific theories. Although Quine admits that some elements of empirical theory are much less likely to be revised than others, he underestimates the asymmetric relations between the hard core and the periphery. It is not just that the periphery is more likely to be revised than the hard core, but rather that the statements of the periphery cannot even be stated, let alone tested, without the hard-core functioning as an a priori in the sense of a necessary precondition. Even if they are all empirical, scientific statements are more than just entrenched, they are constitutive.

In a sense, the difference between Quinean holism and my pragmatic theory of the constitutive elements in science is a matter of emphasis. Quineans emphasize the (ultimate) empirical nature of all knowledge, while I emphasize the constitutive element at the foundation of scientific knowledge. Claiming that there are constitutive elements in scientific theories does not mean, however, that the ultimate grounding of science is a priori. In my view, physical science as a whole remains empirical but only in a very abstract sense that does not tell us much about science. By simply granting that such grounding is empirical, we can set aside the issue of the ultimate justification of the fundamental principles of science and focus attention instead on how these elements of scientific theories function, which is a much more significant question concerning the structure of science. Although I am willing to concede to Quine that the constitutive elements in science are ultimately empirical, the picture of knowledge developed here is still very different from that developed in Quinean holism in that categories of knowledge can be differentiated. Holism, even with a hard core and a maxim of minimal mutilation does not capture the asymmetry that exists between the periphery and the core, when we consider the core as constitutive.

Nevertheless, I see no point in arguing with a reformed Quinean and would rather welcome them on board as a fellow traveler. I still think that Friedman has done a real service by emphasizing the constitutive aspect of the former a priori and that this gives us a more accurate picture of science than Quine's allowed. Quine's viewpoint provides no framework for understanding conceptual change in science, one could argue, because he does not differentiate the constitutive elements enough from the rest of our beliefs and therefore all change seems to be empirical, even if some changes are broader than others.

FRIEDMAN'S NEO-KANTIAN VIEW

I am resisting Friedman's account of the dynamic a priori because he has not made a strong enough case for the positive program that he adopts, looking to philosophy as an independent discipline that provides continuity for the sciences (Friedman 2001, 66). It is possible that the constitutive elements of science function as a priori knowledge but are conventions or even still (ultimately) empirical. To argue for his neo-Kantian a priori, Friedman needs first to distinguish the constitutive elements from the rest of science and then to argue that they cannot be empirically grounded. Even if we leave the question of the status of mathematics to philosophers of mathematics, I am troubled by the rest of the constitutive principles Friedman discusses. Where do we draw the line between the empirical and the constitutive? Pure mathematics versus applied mathematics? Mathematical versus empirical parts of a theory? All of these distinctions are difficult to maintain, given that the constitutive elements themselves have a mixed status. They can be considered to be empirical or to be the a priori foundation of a scientific theory.

Even though I agree with Friedman that the constitutive elements of a physical theory are logically prior to the empirical elements, I do not think that this justifies their being independent in any way. Indeed, I think that we cannot draw a permanent distinction between the constitutive and empirical elements of a scientific theory. As we will see, Pap emphasized that the roles that a particular principle plays in a scientific theory can change, either in time or in a different context. It is not as if we had constitutive elements of theories cordoned off from the rest of science.

Friedman is also on record in opposition to naturalism, yet I find myself to be sympathetic to a modest version of this doctrine. He may simply mean to underline the fact that both Carnap's and Quine's versions of scientism have failed, given that philosophy of science cannot be reduced to logic or to behaviorist psychology. However, the failure of both of these programs does not imply that all versions of naturalism fail. In keeping with naturalism, I also resist Friedman's call for a special role for philosophy. Scientists as well as philosophers need to be aware of the constitutive elements of scientific theories, since philosophers have no claim to having a special ability of being able to point out these elements. I take it that Friedman and I would agree that philosophy is a broad discipline that legitimately does far more than the Vienna Circle thought it could, but I see no reason to think that philosophy plays any special role in science. Indeed, it is important to avoid at all costs the notion of "first philosophy" or any role for philosophy as a separate activity that grounds scientific knowledge, a role that becomes superfluous if we relinquish the general worry about the grounding of science (is it wholly empirical or are there some a priori parts?) and focus instead on how science works, adopting an attitude that goes a step further than Kant by adopting the idea of a relativized a priori and a step beyond the relativized a priori by conceding that even the constitutive elements of physical theory are not independent.

Two things bother me about a priori knowledge as traditionally conceived. First, I am a thoroughgoing fallibilist, and to make sense of that, we need to have everything in our theory be either empirical or conventional. Second, I never found anything like a natural light or rational intuition to be convincing. I have always been sympathetic with the Logical Empiricists for their rejection of rational intuition, which again means that everything must be either empirical or conventional. My task is to make sense of such a position.

Indeed, we may ask, "How could the constitutive elements of a physical theory *not* be empirical?" Surely Newton's mechanics is a physical theory that has empirical consequences. If the theory cannot even be stated without the laws and without the mathematics that are necessary for them, then these would seem to be empirical as well. One way to avoid this conclusion is to claim that the constitutive parts are merely analytic, in which case we are back to defending Carnap from Quine's critique, arguing that they are in some other way separable from the physical theory, for example, by arguing that the pure mathematics refers to one kind of object and the applied mathematics to another. But this cannot be considered a solution, given that it merely pushes the issues back another step to the question of the epistemological status of applied mathematics. Applied mathematics is surely an element of physical theory. The key point is that the constitutive elements of the theory do not fit comfortably as either analytic or synthetic, or empirical or a priori. The fact that a theory has constitutive elements does not imply that it is a priori in the old-fashioned sense, as Kant thought, or even that those elements are nonempirical. The elements can be considered empirical, in the sense that they are part of an empirical theory. The constitutive parts of a theory are those that are only empirical when considered as parts of an empirical whole. Rather than consider the constitutive elements of a theory as requiring a special a priori justification, they can be considered to be justified by the role they play in the theory as a whole.

SOME CLARIFICATION IN RESPONSE TO CRITICS

In order to further clarify my position, I think that it will be valuable to respond to recent articles that call into question the viability of a theory of the constitutive elements in science. The problem raised is the possibility of there being a position that is in between the classical notion of the synthetic a priori as certain and unrevisable, on one hand, and empiricist view that there is no synthetic a priori and that all nonanalytic statements are fallible, revisable and empirical, on the other.

Psillos and Christopoulou (2009) set out these issues clearly and, as I have done, they follow Pap in taking Poincaré's conventionalism as a plausible foundation for a relativized, functional a priori. They believe, however, that there may be no viable middle ground between classical rationalism and classical empiricism. In discussing possible intermediate positions, they reach the following conclusion:

> So we have reached an impasse. If we follow the Poincaréan conception of conventions and develop it in terms of implicit definitions, we have a substantive conception of the constitutive a priori in science—where substantive principles constitute the objects of knowledge. But it is far from clear that we have succeeded in separating the a priori from the empirical. If we follow the Carnapian conception if implicit definitions, we secure a place for the constitutive a priori in science, but it is a rather anemic one. (2009, 220)

The main issue that I need to address is how the conventionalist path to the relativized a priori is any different from empiricism, that is, the first alternative listed above. Specifically, Psillos and Christopoulou ask for a distinguishing mark of a priori (or constitutive) statements (2009, 207). They want something like necessary and sufficient conditions for a statement being functionally a priori or constitutive; an essence, that is, what makes a statement a priori or what explains why it is so. I do not think that we need to worry about meeting this demand for the following reasons: First, we can, of course, say that being a necessary precondition to further inquiry is a criterion for being constitutive or functionally a priori, so it is not as though we have no idea what we are talking about. However, we cannot expect to have necessary and sufficient conditions that pick out all and only the functionally a priori statements. As Pap emphasizes, statements can play different roles at different times and in different contexts. Of course, there will be nothing about a statement that marks it off as permanently functionally a priori (indeed, that is an oxymoron) or as always constitutive. Whether something is functionally a priori is a question of context, not essence. Furthermore, there are degrees of a priorisity, not a strict dichotomy, so many statements will have a mixed status, which explains why Psillos and Christopoulou find that there is no strict boundary between what is a priori and what is not (2009, 208). I would say that this mixed status is precisely what makes some principles of physical science so interesting, given that they seem to be empirical and yet they can play the role of constitutive a priori principles in the context of a physical theory.

Second, we do not need to be able to distinguish a priori statements from empirical ones in advance of studying particular sciences, as Psillos and Christopoulou imply. The point here is to understand science in fine detail, not simply to classify statements as a priori or empirical. We cannot understand the role that a statement is playing before doing an analysis of the science, seeing what is presupposed at a given time and place. The real issue that I am interested in and think that we should be studying is how science works. If we were still debating rationalism versus empiricism and I have to choose, then yes I am an empiricist, but it seems to me that this debate was over by the late eighteenth century when Kant reframed the issues.

Another article that takes the line that there is no viable position between the a priori of classical rationalism and classical empiricism is by Michael Shaffer (2011). He uses this line of argument to question Michael Friedman's dynamic a priori and criticizes my account to the functional a priori as well. Shaffer is quite right when he says that I "stubbornly resist accepting that the view is only nominally different than Quine's" (2011, 202n12), but I reject the implication that I do not have an argument for my position. I have already argued that my position is quite different from Quine's because his is based on entrenchment, not constitution. As I explicitly say in my 2003 article, I side with Quine on many issues (Stump 2003, 1158). However, I also point out that the principles that we are talking about are only empirical in a very tenuous sense and that Quine does not answer the question that I am interested in, namely, that some statements are necessary preconditions to the existence of a particular natural science. If you look at only the epistemological question of whether a statement is empirical or a priori, you get a flattening out of all distinctions between the multiple points of view that have been taken on the constitutive elements in science, simply because you are ignoring all other questions. If you want a nuanced and accurate account of science, then you must consider what is a precondition for something else, what roles various claims are playing at a given time and context, and how these change over time. Only an epistemologist, not a philosopher of science, would stop inquiry at the point of saying that every substantive claim is empirical.

In his critique, Shaffer also assumes that conventions are not justified in any way (2011, 202) which is surely not correct. After all, Poincaré says that conventions can be distinguished by their convenience and that they are informed by experience, and Carnap makes conventional choices part of what he calls pragmatics. I myself argue that the general success of a physical theory, as well as conceptual argument, justifies the choice of constitutive elements in science. It is true that some critics of conventionalism worry that conventions are arbitrary or not rational enough, but that is not actually true of conventionalism as it has developed. Furthermore, it is quite unfair to try to link Friedman to this view, since he has not used the language of conventionalism in developing his view of the dynamic a priori.

SIGNIFICANCE OF THE CONSTITUTIVE ELEMENT IN SCIENCE

What is the upshot of all these theories of the former a priori? The theory of the constitutive elements in science that I am defending is a version of the pragmatic theory of the a priori, different from C. I. Lewis's view, but still a pragmatist answer to the question of the human element in objective knowledge. As Lewis notes, pragmatists are often accused of waffling between a view that holds that human knowledge is objective and a view that holds that human knowledge is subjective, at least in the sense that it is in part created by human action.

> Pragmatism has sometimes been charged with oscillating between two contrary notions: the one, that experience is "through and through malleable to our purpose"; the other, that facts are "hard" and uncreated by the mind. We here offer a mediating conception: through all our knowledge runs the element of the a priori, which is indeed malleable to our purpose and responsive to our need. But throughout, there is also that other element of experience which is "hard," "independent," and unalterable to our will. ([1923] 1970, 239)

The element of knowledge that is constitutive and that had formerly been considered to be fixed is indeed malleable, while the empirical element of knowledge is determined, once the a priori element has been (temporarily) fixed.

One way to look at how we can have both a human element and an objective element in knowledge is to consider the questions that are asked in developing a scientific theory. There are no questions in nature without human beings to ask them; they make up part of the human element in knowledge. The questions that we ask, the demands for explanation, are all a result of human interest, but once the question is asked, there are better and worse answers or explanations in an objective sense. Of course, there are good and bad questions in an objective sense too. Good questions lead to fruitful theories as answers, whereas bad questions do not. The deepest revolutions in the history of science have involved good questions that had good answers, before a new conception changed the entire landscape. These are the conceptual changes that I hope can be explicated with a theory of constitutive elements of science. I begin with the development of non-Euclidean geometries, which deeply influenced philosophers of science and made possible alternative conceptions of space that led the way to revolutionary changes in physics.

NOTES

1. There is a helpful discussion of Friedman's use of the term *constitutive* in Richardson (2002) and of Ryckman's usage in van Fraassen (2007).
2. There is some controversy over whether Newton himself uses the calculus in the *Principia*. Niccoló Guicciardini (1998, 1999) has made a compelling case that he did. Of course, both the calculus and Newtonian physics developed and changed quite a bit over time. What can be said uncontroversially is that Newtonian physics as it was developed requires the calculus as it was developed. See also Friedman (2001, 39n45).
3. See Toulmin (1961) for a detailed account of this example, which I consider again in Chapter 6.
4. Drafts are available at Norton's website, www.pitt.edu/~jdnorton/homepage/ cv.html#manuscripts.

5. A common reply to skeptical arguments based on holism is that it is a mere philosophical abstraction to say that one can always imagine an alternative and that furthermore an actual alternative is necessary before one can legitimately question a theory or an experimental result (Austin 1961, 45, cited in Stroud 1984, 45).

BIBLIOGRAPHY

Austin, J. L. 1961. "Other Minds." In *Philosophical Papers*, edited by J. O. Urmson and G. J. Warnock. Oxford: University of Oxford Press, 76–116.

Dayton, Eric. 2006. "Clarence Irving Lewis (1883–1964)." *Internet Encyclopedia of Philosophy.* www.utm.edu/research/iep/l/lewisci.htm.

Friedman, Michael. 2001. *Dynamics of Reason: The 1999 Kant Lectures at Stanford University.* Stanford, CA: CSLI Publications.

———. 2002. "Kant, Kuhn, and the Rationality of Science." *Philosophy of Science* 69 (2): 171–190.

———. 2003a. "Transcendental Philosophy and Mathematical Physics." *Studies in History and Philosophy of Science Part A* 34 (1): 29–43.

———. 2003b. "Kuhn and Logical Empiricism." In *Thomas Kuhn*, edited by T. Nickles. Cambridge: Cambridge University Press, 19–44.

———. 2004. "Philosophy as Dynamic Reason: The Idea of a Scientific Philosophy." In *What Philosophy Is: Contemporary Philosophy in Action*, edited by H. Carel and D. Gamez. London: Continuum, 73–96.

———. 2005a. "Ernst Cassirer and Contemporary Philosophy of Science." *Angelaki: Journal of the Theoretical Humanities* 10 (1): 119–128.

———. 2005b. "Transcendental Philosophy and Twentieth Century Physics." *Philosophy Today* 49 (5 [Suppl.]): 23–29.

———. 2006. "Carnap and Quine: Twentieth-Century Echoes of Kant and Hume." *Philosophical Topics* 34: 35–58.

———. 2008a. "Einstein, Kant and the A Priori." In *Kant and Philosophy of Science Today*, edited by M. Massimi. Cambridge: Cambridge University Press, 95–112.

———. 2008b. "Ernst Cassirer and Thomas Kuhn: The Neo-Kantian tradition in History and Philosophy of Science." *Philosophical Forum* 39 (2): 239–252.

———. 2009. "Einstein, Kant and the Relativized A Priori." In *Constituting Objectivity: Transcendental Perspectives on Modern Physics*, edited by M. Bitbol, P. Kerszberg, and J. Petitot. Berlin and New York: Springer, 253–267.

———. 2010a. "A Post-Kuhnian Approach to the History and Philosophy of Science." *The Monist* 93 (4): 497–517.

———. 2010b. "Synthetic History Reconsidered." In *Discourse on a New Method: Reinvigorating the Marriage of History and Philosophy of Science*, edited by M. Domski and M. Dickson. Chicago and La Salle, IL: Open Court, 571–813.

———. 2011. "Extending the Dynamics of Reason." *Erkenntnis* 75 (3): 431–444.

———. 2012. "Reconsidering the Dynamics of Reason: Response to Ferrari, Mormann, Nordmann, and Uebel." *Studies in History and Philosophy of Science Part A* 43 (1): 47–53.

Guicciardini, Niccoló. 1998. "Did Newton Use His Calculus in the Principia?" *Centaurus* 40 (3–4): 303–344.

———. 1999. *Reading the Principia: The Debate on Newton's Mathematical Methods for Natural Philosophy from 1687 to 1736.* Cambridge: Cambridge University Press.

Hanson, Norwood Russell. 1958. *Patterns of Discovery: An Inquiry into the Conceptual Foundations of Science.* Cambridge: Cambridge University Press.

Kuhn, Thomas S. 1962. *The Structure of Scientific Revolutions*. Chicago: University of Chicago Press.

Lewis, Clarence Irving. (1923) 1970. "A Pragmatic Conception of the A Priori." In *Collected Papers of Clarence Irving Lewis*, edited by J. D. Goheen and J. John L. Mothershead. Stanford, CA: Stanford University Press, 231–239.

Manchak, John Byron. 2009. "Can We Know the Global Structure of Spacetime?" *Studies in History and Philosophy of Science Part B: Studies in History and Philosophy of Modern Physics* **40** (1): 53–56.

Norton, John D. 2011. "Observationally Indistinguishable Spacetimes: A Challenge for Any Inductivist." In *Philosophy of Science Matters: The Philosophy of Peter Achinstein*, edited by G. Morgan. Oxford: Oxford University Press, 164–176.

Oberdan, Thomas. 2009. "Geometry, Convention, and the Relativized A Priori: The Schlick—Reichenbach Correspondence." In *Stationen. Dem Philosophen und Physiker Moritz Schlick zum 125. Geburtstag*, edited by F. Stadler, H. J. Wendel, and E. Glassner. Vienna: Springer, 186–211.

Pap, Arthur. 1946. *The A Priori in Physical Theory*. New York: King's Crown Press.

Psillos, Stathis, and Demetra Christopoulou. 2009. "The A Priori: Between Conventions and Implicit Definitions." In *The A Priori and its Role in Philosophy*, edited by N. Kompa, C. Nimtz, and C. Suhm. Paderborn, Germany: Mentis Verlag GmbH, 205–220.

Quine, Willard Van Orman. 1992. *The Pursuit of Truth*. Rev. ed. Cambridge, MA: Harvard University Press.

———. 1995. *From Stimulus to Science*. Cambridge, MA: Harvard University Press.

Reichenbach, Hans. (1920) 1965. *The Theory of Relativity and A Priori Knowledge*. Berkeley: University of California Press.

Richardson, Alan W. 2002. "Narrating the History of Reason Itself: Friedman, Kuhn, and a Constitutive A Priori for the Twenty-First Century." *Perspectives on Science* **10** (3): 253–274.

Shaffer, Michael J. 2011. "The Constitutive A Priori and Epistemic Justification." In *What Place for the A Priori?* edited by M. L. Veber and M. J. Shaffer. Chicago: Open Court, 193–209.

Stöltzner, Michael. 2009. "Can the Principle of Least Action Be Considered a Relativized A Priori?" In *Constituting Objectivity: Transcendental Perspectives on Modern Physics*, edited by M. Bitbol, P. Kerszberg, and J. Petitot. Berlin and New York: Springer, 215–227.

Stroud, Barry. 1984. *The Significance of Philosophical Scepticism*. New York: Oxford University Press.

Stump, David J. 2003. "Defending Conventions as Functionally A Priori Knowledge" *Philosophy of Science* **20**: 1149–1160.

Toulmin, Stephen. 1961. *Foresight and Understanding: An Enquiry into the Aims of Science*. Bloomington: Indiana University Press.

Tsou, Jonathan. 2010. "Putnam's Account of Apriority and Scientific Change: Its Historical and Contemporary Interest." *Synthese* **176** (3): 429–445.

van Fraassen, Bas C. 2007. "The Constitutive A Priori: Review of Ryckman, *The Reign of Relativity*." *Metascience* **16** (3): 407–419.

2 Reinventing Geometry as a Formal Science

The formal conception of geometry and the historical accounts that inform it were extremely influential in twentieth-century philosophy of science. A formal conception of geometry entailed a radical shift in the conception of geometry as a science. Rather than being taken to be true of the world, pure geometry could now be taken to be a merely convenient tool that does not express any statement about the world. This demoting of geometry had a significant effect on what had been taken to be a priori knowledge, given that for Kant, space and time are given in intuition, forming the basis of geometry and arithmetic, respectively. A formal conception of geometry is, in a logical sense, a necessary precondition to the reconsideration of what had been an a priori science. The development of non-Euclidean geometries had a profound effect on philosophers such as Poincaré, who in turn stimulated rethinking of the role of constitutive elements in science through his conventionalism.

Geometry was always the most rigorous and the most formal of the sciences that describe the world, but by the beginning of the twentieth century, it was necessary to reconceive geometry of as an abstract science that does not describe the way the world is. According to what has been called the traditional, or Aristotelian view, a science must have a well-defined subject matter and express (or at least aim at finding) truths about that subject matter. The objects to be studied should be determined at the beginning of the inquiry. Geometry is the science of space, or more narrowly, it is the science of points, straight lines, angles, circles, planes, and so on. By the end of the nineteenth century, a formal view of mathematics was taking hold in which geometry was viewed as a pure uninterpreted axiomatic system that does not express truths. The primitive terms in geometry were taken to refer to nothing at all; thus, they could be explicitly redefined to refer to any set of objects that maintained the abstract relationships stated in the axioms. A formal conception of mathematics can be thought of as a dogma of the twentieth century. For example, in 1946 the physicist Percy Bridgeman wrote, "It is the merely truism, evident at once to unsophisticated observation, that mathematics is a human invention" (cited in Kline 1980, 325). One influential story of how we came to this twentieth century view starts in nineteenth century geometry.

EUCLIDEAN GEOMETRY AND THE PARALLEL POSTULATE

Formal systems are axiomatic and deductive. They are deductive because in a formal system, one proves what is true using logical rules that have been spelled out in advance. The truths of the formal system are theorems and we know that they are true because they can be proved on the assumption of more basic truths. However, we cannot prove everything by this method because there will come a point where there is nothing more basic on which we could base a proof of a theorem. All formal systems therefore assume certain basic truths without proof, and these are the axioms. A major part of formalizing knowledge is making explicit all of the assumptions that one uses, that is, listing the axioms that one takes to be true at the beginning. A similar process occurs with definitions of basic terms. In a formal system, one attempts to explicitly define as many of the special words that are used as possible. Again, we cannot define everything, so we come to some basic terms that are not defined explicitly in the formal system; these are the primitive terms.

Euclidean geometry is the geometry that most of us have studied in high school. To keep things simple, I discuss only plane geometry (two-dimensional) and consider only one alternative to Euclid's geometry. The focus of a long-standing discussion in geometry is parallel lines, that is, lines in the same plane that do not intersect. In Euclid's version, the parallel postulate (or axiom) says that if a line forms an acute angle to the perpendicular of a given line, it will eventually intersect with the given line.

Even some early commentators found this formulation of the postulate troubling. Perhaps part of the problem is that the intersection of the two lines may occur off the picture. We cannot see that the lines intersect, yet we know that they must. For two thousand years, mathematicians attempted to replace Euclid's parallel postulate with something clearer, for example, Playfair's popular version of the postulate, that through a point off a given line one can draw exactly one parallel line, but they always found that the formulation that was used was equivalent to Euclid's original postulate. In precise terms, this means that one could start with Euclid's parallel postulate and the other axioms of geometry and prove the replacement as a

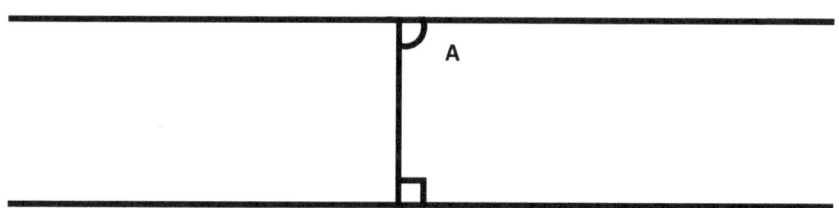

Figure 2.1   Euclid's Parallel Postulate
Source: Drawing by David J. Stump.

theorem, and one could also start with the replacement and the other axioms of geometry and prove Euclid's original axiom. Each time a proof of the parallel postulate was offered, it was shown that some implicit or explicit assumption which is equivalent to the parallel postulate was being used in the proof. Surprisingly, many assumptions that seemed to have nothing to do with parallel lines and that were proposed as harmless assumptions that would allow one to prove Euclid's parallel postulate were also shown to be equivalent to the postulate. For example, the assumption that there are similar triangles (triangles with the same angles but with sides of different length) is equivalent to Euclid's parallel postulate. The process of proposing proofs of Euclid's parallel postulate and showing that they always required assumptions that are equivalent to the postulate taught mathematicians a lot about what is equivalent and what is not, that is, they learned a lot about geometry. Since the postulate and the replacement are equivalent or interchangeable, it is clear that if mathematicians were trying to eliminate the assumption of the postulate, they were not getting anywhere. Some explicit statement which is equivalent to the parallel postulate was necessary to proceed with the task of proving the theorems of geometry.

In the early eighteenth century, Girolamo Saccheri (1733) tried a new indirect approach to proving Euclid's parallel postulate, reductio ad absurdum. In order to prove something by reductio ad absurdum, one first assumes the opposite of what one wants to prove and then shows that this assumption leads to a contradiction. This shows that the original assumption of the opposite of what one wants to prove is false, so that what you want to prove is true. Saccheri started with Figure 2.1 and thought of three cases. If the angle is a right angle, there will be only one parallel. If the angle is obtuse, he proved that the assumption that the line remains parallel (i.e., that it does not intersect the lower line) leads to a contradiction. However, in the case where the angle is acute, Saccheri had some problems. He assumed that the upper line did not meet the lower line, he proved many strange theorems, and he thought he had a contradiction. Actually, he did not find a formal contradiction by this method, something that I will return to in a moment. Let us consider one of the many strange theorems that can be proved on the assumption that the upper line does not intersect the lower line when the angle in Figure 2.1 is acute. In the first place, it should be clear that there is more than one parallel line through a given point not on the line; indeed, there are an infinite number of parallels to a given line. Furthermore, to take an extremely counterintuitive case, given any two lines that are perpendicular, it is always possible to find a third line that is parallel to both.

Notice that the question of whether the third line will intersect one of the two perpendicular lines is precisely the same as Euclid's original parallel postulate; that is, will straight lines that intersect a perpendicular at an acute angle intersect the given line? If the parallel postulate is false, then the strange theorem is true, since there are straight lines that intersect the perpendicular at an acute angle yet never meet the line at the base.

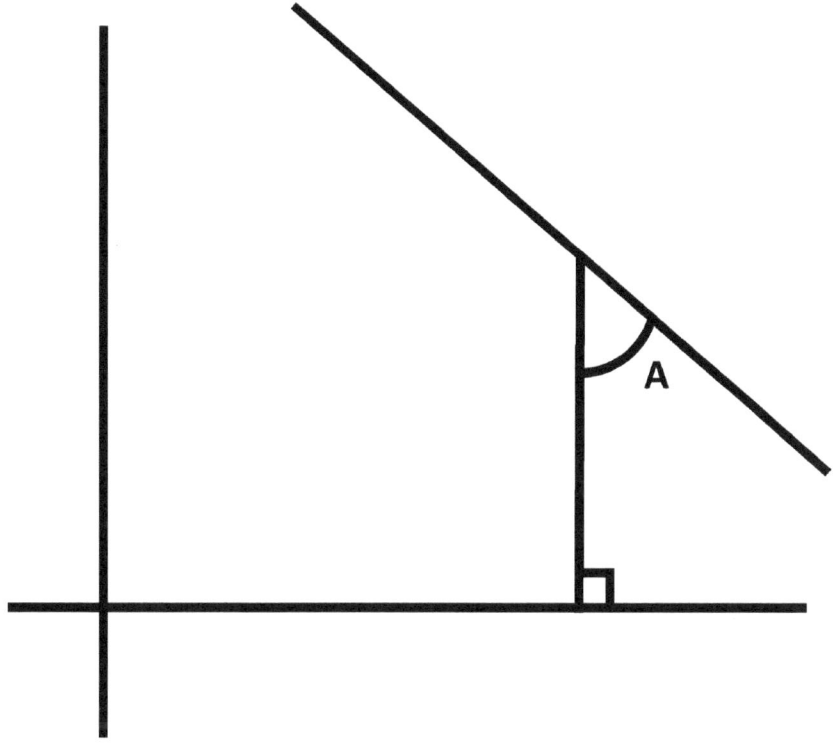

Figure 2.2   A Theorem of Hyperbolic Geometry
Source: Drawing by David J. Stump.

Around 1830 two mathematicians, N.I. Lobachevskii and J. Bolyai, working independently each published works in which they claimed that this geometry was consistent; that is, they denied the claim that the assumption that a line forming an acute angle may not intersect leads to a contradiction. In 1869, Beltrami gave a model; that is, he showed how to interpret Bolyai and Lobachevskii's new geometry (hereafter, hyperbolic geometry) as part of a Euclidean surface.

Following the work of Gauss who developed a way of studying the geometry on surfaces, hyperbolic geometry can be considered a geometry on a saddle-shaped surface, that is, one that curves in two different directions. Klein provided a different model of hyperbolic geometry, in which the straight lines of hyperbolic geometry are considered as the open-ended line segments within a disk. Taking these segments as lines, we can easily see that it is possible to draw more than one parallel (remember, lines that do not intersect) to a giving line through a point off of the line and that our strange theorem that given two lines that are perpendicular, it is always possible to draw a line that is parallel to both is obviously true.

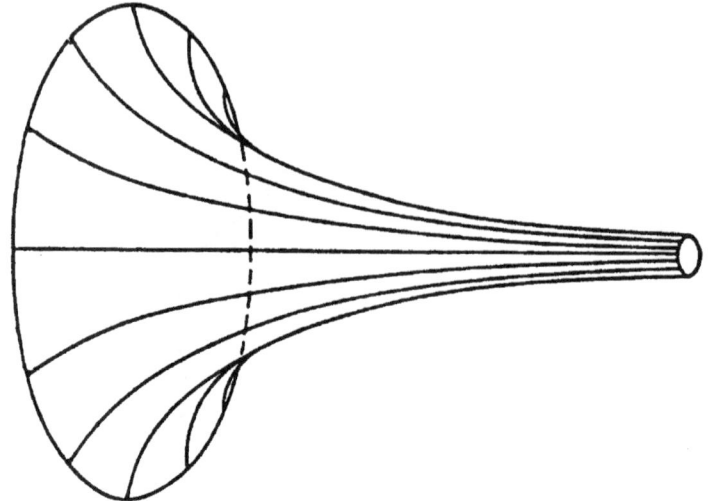

Figure 2.3   Beltrami's Model: The Pseudosphere

Source: "Figure 1.1: The Pseudosphere," in John Stillwell, *Sources of Hyperbolic Geometry* (Providence, RI: American Mathematical Society, 1996), 1, by permission of the American Mathematical Society.

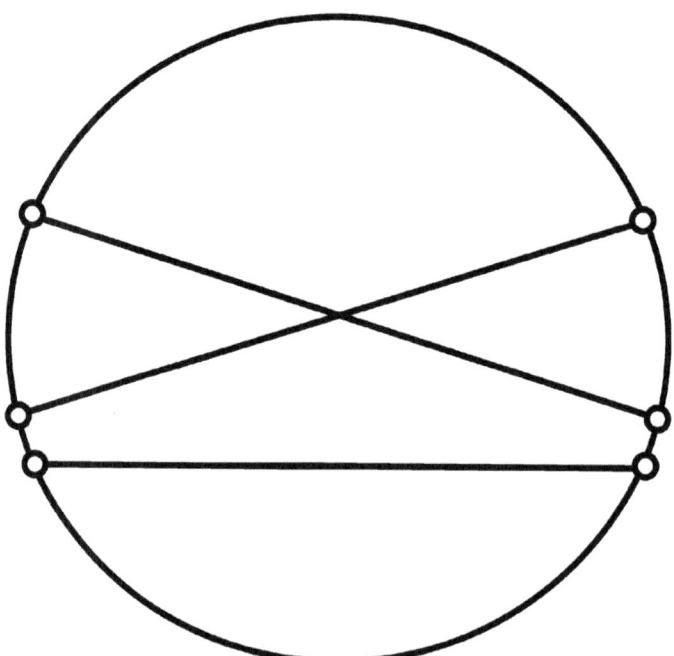

Figure 2.4   The Multiple Parallels in Klein's Model

Source: Drawing by David J. Stump.

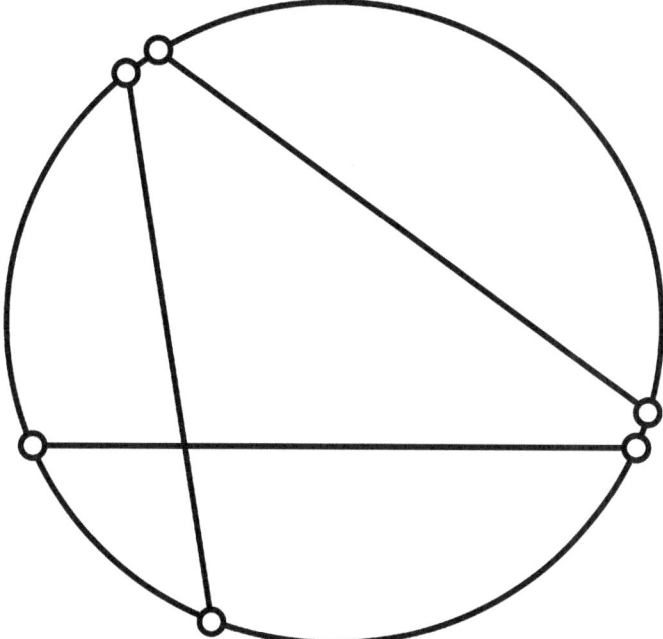

Figure 2.5   Theorem of Hyperbolic Geometry in Klein's Model
Source: Drawing by David J. Stump.

These models were taken to show that hyperbolic geometry is consistent. That is, we can see that it is possible to have more than one parallel line because the models show us that there are surfaces or analytical constructs where it is true that there is more than one parallel. Therefore, the assumption that a line forming an acute angle to a perpendicular never intersects the given line does not lead to a contradiction, since there are objects in Euclidean geometry that exhibit these properties. Of course, taking these models to represent hyperbolic geometry assumes that it is possible to assign new meanings to the term *straight line* in hyperbolic geometry. The debate over the formalism of geometry at the turn of the century is a debate over why we now consider it legitimate to redefine the terms of a formal system in this manner; that is why our preconceived idea of what straight lines is no longer was taken to rule out the possibility of there being more than one parallel line. The new conception of geometry allows for a change in what had been considered a priori knowledge.

LOGIC AND GEOMETRY

The development of non-Euclidean geometry has often been seen as the result of a growing movement toward more rigor, more explicit definitions, more emphasis on proof, and the ferreting out of implicit assumptions, ending with modern logic and mathematical formalism (Bonola [1912] 1955; Coolidge [1940] 1963; Kline 1972). In these accounts, concern over foundational issues is thought to have led to the development of non-Euclidean geometries. For example, the famous two-thousand-year-long discussion of the parallel postulate is seen as an attempt to prove the postulate rigorously, implying that geometers were uncertain whether Euclid's postulate was true. In his influential history of non-Euclidean geometries, Bonola says that

> [e]ven the earliest commentators on Euclid's text held that Postulate V. was not sufficiently evident to be accepted without proof, and they attempted to deduce it as a consequence of other propositions. (1955, 3)

Morris Kline has correctly noted that often the effort in discussions of the parallel postulate was to find the simplest statement among those which are intuitively true or to see whether the assumption of the postulate is really necessary, rather than to doubt its truth (1980, 78). Another frequently heard charge, that mathematicians were fooled over and over again into using implicit assumptions that are equivalent to the parallel postulate while attempting to prove it is misleading at best, since many mathematicians were explicitly replacing the parallel postulate with a different assumption. Some contemporary writers have gone so far as to claim that geometers came to a realization that they were dealing with a purely formal system and that terms that seemed meaningful and axioms which seemed assertions actually are not (Freudenthal 1957; Steiner 1964; Contro 1976; Trudeau 1987, 162). Such claims are clearly false, since a formal view of mathematics did not exist until long after non-Euclidean geometries were developed and accepted.

   All of the historical accounts agree that the development of non-Euclidean geometries falls neatly into historical periods even if they do not see, as Jeremy Gray has emphasized, that different issues were addressed in each period ([1979] 1989, 168–169). I follow Gray in calling this the standard account. In 1830, Bolyai and Lobachevskii were responding to the traditional disputes over the parallel postulate, but the work of Bolyai and Lobachevskii received little attention. In late 1860s, Beltrami and Klein provided interpretations of Bolyai-Lobachevskii geometry (hyperbolic geometry), although this method was quickly superseded by Klein's Erlangen Program, which organized various geometries under projective geometry and, when considered algebraically, as the study of groups of transformations. By the 1870s, after the works of Beltrami and Klein, non-Euclidean geometries textbooks and general expositions of non-Euclidean geometries appeared. Ironically, there are no critiques of non-Euclidean geometries until after 1870, immediately after the work of Riemann, Beltrami, and Klein had finally put hyperbolic geometry on a firm foundation (see Sommerville [1911] 1970). It was not until Riemann (see Nowak 1989) and Helmholtz gave a philosophical interpretation of the existence of non-Euclidean geometries, pointedly argu- ing that their existence refuted Kant's philosophy, that non-Euclidean geom- etries were widely discussed and debated. At the turn of the century, we get debates over foundational issues. Hilbert and Frege and Poincaré and Russell debated the interpretation of primitive terms and related issues on the nature of definitions and axioms, the role of intuition in mathematics, which by then was suspect, and Peano and Hilbert and Frege wrote new axiom systems for geometry. By that time formal axiomatics was already well developed (marked by the International Mathematical and Philo- sophical Congresses in Paris in 1900).

The debate over non-Euclidean geometries after 1870 is not a founda- tional crisis as presented in the standard account. Indeed, the consistency of hyperbolic geometry and the nonprovability of the parallel postulate was already accepted before the work of Riemann, Beltrami, and Klein, even though everyone admits that the work of Bolyai and Lobachevskii is woefully inadequate from a logical point of view. Bolyai and Lobachevskii assumed the acute angle hypothesis and derived consequences, and they developed a non-Euclidean geometry by proving theorems; however, they had little to say about consistency. For example, in his first paper Lobachevskii claimed that his geometry was consistent because he obtained its trigonometric for- mulas from the corresponding formulas of spherical geometry by multiply- ing the sides of a triangle by an imaginary unit. All that is proved is that these formulas are consequences of the assumptions of hyperbolic geometry; what is required is to show that all the formulas of hyperbolic geometry are consequences of some other geometry, a proof which Lobachevskii did attempt in a latter work without arriving at a complete proof (Rosenfeld 1988, 227–228).

The argument given in textbooks of the time was on even less of a sound logical footing. It was simply stated that the attempt to prove the paral- lel postulate had gone on too long and that the repeated lack of success shows that it is impossible (this is a traditional argument, often mentioned in the French literature, which goes back to d'Alembert and Klügel). Here is a statement on the status of the parallel postulate before the work of Riemann, Beltrami, and Klein had appeared from Jules Hoüel, the French mathematician who was as responsible as Riemann and Helmholtz for making the existence of non-Euclidean geometries known:

> Research which is already old, but which has passed unnoticed until recently, has put beyond doubt that the 11th axiom of Euclid (our 4th axiom) cannot be deduced from the preceding axioms. This dis- covery was made around 1829 by two geometers, Lobachevskii and J. Bolyai, . . . (Hoüel 1867, 72)

Manning's textbook of 1901 takes exactly the same point of view, and it is most striking that he does not mention the models of hyperbolic geometry that had been given by Beltrami, Klein, and Poincaré (Manning 1901). Of course, we could simply say that Hoüel and Manning, among others, made mistakes; that they did express themselves clearly; and that they did not formulate the arguments that were available to them. I do not accept such reasoning; what is important here is that there is no evidence that Riemann, Beltrami, or Klein were responding to a crisis in the foundations of geometry. Quite to the contrary, geometers seemed to have accepted the consistency of hyperbolic geometry and the impossibility of proving the parallel postulate even before the work of Riemann, Beltrami, Klein, and Hilbert had put the non-Euclidean geometries on a sound logical footing.

Therefore, it seems clear that until 1870 at least, something other than logic is behind the acceptance of the consistency of hyperbolic geometry. Geometers in this period did not treat geometry as pure uninterpreted axiomatic systems, and yet they accepted the consistency of hyperbolic geometry. The arguments against hyperbolic geometry, and non-Euclidean geometries generally, fall into two basic categories, one philosophical and one mathematical. The philosophical discussions appeared in response to Helmholtz's famous and widely circulated attack on Kant's theory of geometry and revolve around the issue of which geometry is true and how we know this, which is certainly continuous with earlier philosophical discussion of mathematics and of a priori knowledge in general. The fact that empiricism in geometry had already been championed by Mill, for example, shows that there is more continuity than might be expected in the philosophical debate over Euclid, and in that sense, the revival of empiricism by Riemann and Helmholtz can hardly be seen as revolutionary. A revolution did occur in philosophy when formalism in mathematics took hold, but interestingly, many of the major actors in that movement do not see a connection to geometry. There is no revolution in geometry until stock is taken of what happened, and this is true even in mathematics proper. For example, the development of spherical geometry was not considered revolutionary at all, because it was not seen to conflict with Euclid. In fact, spherical geometry conflicts Euclid every bit as much as hyperbolic geometry (Gray [1979] 1989, 71). It was only when historians started drawing philosophical consequences from the development of geometry in the nineteenth century that the revolution which had already taken place could be seen.

Some of the Kantian replies are quite sophisticated, for example, Charles Renouvier (1889, 1891; see Torretti 1984, 294) argues that the consistency of the non-Euclidean geometries supports Kant's position. After all, Kant claimed that the truths of geometry are not analytic, and if hyperbolic geometry was formally inconsistent, then Euclidean geometry would be analytically true. Renouvier still argues that hyperbolic geometry is inconsistent in the sense that it goes against necessary intuition.

Most of the mathematical arguments against the consistency of hyperbolic geometry seem to amount to new proofs of the parallel postulate by elementary methods. Once a question-begging assumption is made, it is easy to show that hyperbolic geometry is wrong and that Euclidean geometry is right. I can cite articles by Dauge and by Combebiac that appeared in *Mathesis* and *L'Enseignement Mathematique* in the 1890s as examples (Dauge 1896; Combebiac 1903). Neither the philosophical nor the mathematical arguments had any effect on the mathematics community, as far as I have been able to determine. The philosophical debate was largely between philosophers who were defending Kant and physicists such as Helmholtz and Mach. With few exceptions, the mathematical criticisms were developed using very elementary techniques, and these were handled by more sophisticated mathematicians with letters to the editor: one-page refutations of the arguments. Mathematicians using advanced techniques went on developing new results in geometry without bothering to respond to the critics in their major works. The percentage of geometrical works devoted to the theory of parallels declined sharply after 1870 and continued to decline through the last decades of the nineteenth century (Sommerville [1911] 1970). It is interesting to note that there is virtually no discussion of the mathematical works which were critical of non-Euclidean geometries in the secondary literature, including the standard histories. For example, Torretti ([1978] 1984) discusses the philosophical critics but not the mathematical ones.

NAGEL'S ACCOUNT

The points that I have raised earlier cast doubt on the standard account. In a classic and widely cited paper, Nagel reverses the order of the standard account and argues that the development of analytic techniques in geometry provided the means for developing formal axiomatics at the turn of the century, basically by showing that geometric primitives are uninterpreted or reinterpretable and that the abstract relationships given in the axioms remain the same under different interpretations (Nagel [1939] 1979; also see Bourbaki [1969] and Freudenthal [1974], who agree with Nagel). However, it is far from clear how much the development of nineteenth-century geometry influenced the modern conception of logic and mathematics. From the modern perspective of pure mathematics, we can look back and see ideas and methods that fit well with a formal conception of mathematics, but we can also see many ideas and methods that do not. Although I cannot tell the complete history here, I want to justify my claims by criticizing both the standard and Nagel's accounts of the development of non-Euclidean geometry.

Curry's and Wilder's textbooks on the foundations of mathematics make my case very well. In both of these standard texts, the motivation for formalization and the study of the foundations of mathematics is the set theoretical paradoxes, not the discovery of non-Euclidean geometries (Wilder 1952,

53–57; Curry 1963, 3–4). The same type of presentation is found in *Principia Mathematica* (Whitehead and Russell [1910–1913] 1927) although Russell was well aware of the development of non-Euclidean geometries and had written quite a lot about its philosophical implications. Famously and perhaps notoriously, Frege maintained that there was a profound difference in the grounding of geometry and arithmetic (see Bynum 1993). Thus, those studying the foundations of mathematics immediately after the turn of the century did not see a revolution in geometry as the basis for their new work, and to this day, textbook justifications of the study of the foundations of mathematics does not include geometry. By striking contrast, the contemporary undergraduate textbook presentation of geometry does include a heavy dose of formal axiomatic methods, a puzzle I try to explain in the following.

Unfortunately, Nagel has not established any genuine links between the middle of the nineteenth century and the early twentieth. Nagel shows that some ideas about interpretation of primitives in models appear in early analytic work in geometry, connecting ideas without showing concretely that earlier authors were read by those of the later period. There is no doubt that the end point colors what comes before; Nagel claims that what we do now is implicit in the practices of nineteenth-century mathematicians yet, for some reason, they were unable to see that. For example,

> [i]t is true that Poncelet was not conscious of what he had done, and he did not in fact possess an adequate philosophy of symbolism in terms of which he could have viewed his accomplishment in an appropriate light. He would hardly have assented to the view that the task of the pure geometer (or the pure mathematician in general) is the exploration of the mutual interrelation of signs governed by specific rule of operation, irrespective of the "interpretations" or "meanings" which may be assigned to them. Nonetheless, at least on one occasion he was definitely on the road which terminates in such an outlook. (Nagel [1939] 1979, 206)

Ideas develop even if the historical actors involved do not understand what they are doing in Nagel's account. He presents the formalization of mathematics as a natural, perhaps even necessary development:

> That pure geometry is neither a science of quantity nor the science of extension is the conclusion which naturally emerges from the development of geometrical ideas thus far traced. Everything in this history points to the view that the only identifiable subject matter which can be assigned to demonstrative geometry is the interconnections of the symbolic operations whose properties have been formally specified. However, the clear emergence of such a doctrine was hindered for a number of years for a number of reasons. (Nagel [1939] 1979, 222–223)

My argument against Nagel's view is that parts of the story, such as treating primitive terms as uninterpreted, existed before the development of hyperbolic geometry in the mid-nineteenth century and was not applied generally to other mathematical primitives. Nagel needs to show that the techniques and outlooks developed in nineteenth-century mathematics were specifically taken up and used in the development of formal logic. For example, Nagel's claim that the expansion of algebra influenced Boole sounds plausible (Nagel [1935] 1979), although it has been criticized by Sherry (1991), but in geometry, while it is true that geometry was being generalized and transformed, exactly how this effected logic is not clear. Of course, it is true that formal axiomatics provides a powerful framework for proving the consistency of hyperbolic geometry since it provides requisite definitions of the logical independence of axioms and of relative logical consistency; however, formal axiomatics did not exist when the consistency of the non-Euclidean geometries was being accepted.

FOUNDATIONS OF MATHEMATICS

Both the standard account and Nagel's account strongly link the history of the discussion of the parallel postulate and discovery of non-Euclidean geometries with the development of mathematical logic and discussion of the foundations of mathematics. This linkage is also stressed by contemporary geometry textbooks which treat the discussion of the parallel postulate and the consistency of hyperbolic geometry as a historical example of a problem that can be solved with modern axiomatics and logic, with the geometry sometimes getting less treatment than axiomatics (e.g., Meschkowski 1964; Greenberg 1980; Trudeau, 1987; Perry 1992). Nothing is necessary wrong with this approach, as long as it is not claimed to be historical; in fact, it is interesting and I think important to note that the development of non-Euclidean geometries should be taken to be such a good pedagogical topic for introducing axiomatics, although it is equally interesting to note that earlier textbooks developed systems of non-Euclidean geometry without formal proofs of consistency (Manning 1901; Coolidge 1909). My claim is that formal logic and the study of the foundations of mathematics affected the writing of the history of the non-Euclidean geometries and that the narrative that was established has affected pedagogical writing on geometry to this day (see Gray 1987, 57). The narrative history that developed from a formal perspective also profoundly changed the philosophical implications that were drawn from the development of non-Euclidean geometries, although as is typical of philosophical discussions, the implications that should be drawn were strongly contested.

I want to highlight the fact that a formal conception of mathematics allowed a redescription of the development of non-Euclidean geometry and on how this description has become the standard, if whiggish, account. I do not want to browbeat people for being whiggish, but rather to reflect on the role of history in justifying the study of foundations of mathematics and the use of formal logic and axiomatic methods in undergraduate geometry textbooks. It is well known that histories of the sciences are often used to legitimate new sciences or revolutionary changes (see the papers in Graham, Lepenies, and Weingart 1983). The use of the history of the parallel postulate to justify the study of the foundation of mathematics and formal logic in elementary geometry textbooks is surely an example of such legitimation, especially since the original discussion of the foundations of mathematics did not include discussion of geometry as a justification. I want to go further and say that history, as well as logic, helped formalize mathematics. The narrative reconstruction of the development of geometry in the nineteenth century is part of the content of science and just as much part of development of geometry as the mathematical techniques are. By this I mean quite straightforwardly that the history of non-Euclidean geometry is part of what is taught in standard university courses in geometry and that a connection between the history of the discussions of the parallel postulate, the need to formalize axiomatic systems in general, and for rigorous proof in general is part of this standard curriculum.

THE UNITY OF MATHEMATICS

There is another positive side to the standard account that should be mentioned, however, since it shows the power of the study of foundations that occurred at the turn of the century. Diverse mathematical practices could be integrated into a new framework and justified by powerful methods in compelling ways. The consistency of hyperbolic geometry could be proved with new rigor, and all of mathematics could be seen as one large axiomatic system. This new framework was the product of a new interdisciplinary coalition between philosophers and mathematicians, and the process of developing this interdisciplinary coalition is an interesting area to study. Philosophers were apparently quite hostile to non-Euclidean geometry up to the turn of the century, until mathematicians and philosophers started working together on foundations. Analogously, mathematicians did not have much use for philosophers until they started to work together on the foundations of mathematics and on formal logic. The philosophy of mathematics was created as a specialization in philosophy in the process and became part of a movement towards a so-called scientific philosophy with the later development of Logical Empiricism.

I look at the two early influential histories of the non-Euclidean geometries, that of Stäckel and Engel, published in German in 1895, and that of Bonola, published in Italian in 1906, expanded and translated into German in 1908, reprinted 1919 and 1921, translated into English in 1912, reprinted in 1938 and 1955, and translated into Spanish in 1945. Although Stäckel and Engel did most of the original historical work, Bonola's book was much more widely disseminated. A statement from E. Carruccio, if taken literally and as if he was speaking for the whole mathematical community, instead of how it was intended, could serve to highlight the impact of Bonola's history: "when nothing is said to the contrary, we refer always to Bonola" (1964, 251). What was the motivation for writing, translating and publishing the history of non-Euclidean geometries? The motive was not to convince mathematicians that non-Euclidean geometries are consistent; that issue had been settled by 1870. There seem to have been two main motivations: one is to convince philosophers that non-Euclidean geometries are consistent and to draw philosophical lessons from the development of non-Euclidean geometries, and the other is to give teachers of mathematics new tools for pedagogy. Both of these motivations require the development of non-Euclidean geometries to be described in a form that will be useful, and it appears that describing the developments as a revolution that occurred in response to a crisis was just what was needed to show the limits of the rational intuition that classical mathematicians and philosophers had used, and to convince students that they had better learn how to express themselves in symbolic terms and to formulate rigorous proofs.

In his introduction, Stäckel says that he wrote his history partly in a conscious attempt to convince philosophers:

> We have already changed the minds of mathematicians, so we would like to bring our book to the attention of philosophers, then the theory of parallels will stand in close connection with similar foundational problems of philosophy, touching on, as Gauss expressed himself, the point in question in Metaphysics. (Stäckel and Engel 1895, vi)

Bonola, on the other hand, says very little about his explicit motivation for writing his history. He does say that his history of the non-Euclidean geometries was written in response to "the interest felt in the critical and historical exposition of the principles of various sciences" (Bonola [1912] 1955, vii) and titled his book *Non-Euclidean Geometry: A Critical and Historical Account of its Development*. Mach published his *Science of Mechanics* with the same subtitle that refers to historical criticism, and indeed, references to the critical historical method in writings about science is ubiquitous in the late nineteenth century. Turning to *Baldwin's Dictionary of Philosophy and Psychology* for some idea of what such a phrase might mean in the period, I found the following as the definition of historical criticism (with reference to Goethe, the Schlegels, Schiller, Hegel, and Coleridge): "[the work of art's] interpretation and expression of the life and interests of the people and age in which it arose" (Baldwin [1901–1905] 1960, 246). Thus, at least part of the idea of a critical historical account is that the new geometry should be understood in the vocabulary of the contemporary figures who developed it. Stäckel also claims that to understand the essence of non-Euclidean geometry, one must understand the work of Lobachevskii and Bolyai without importing later mathematical ideas, although it is unclear how widely felt the need for historical understanding was at the turn of the century or how much these authors hold to their own standards.

The preface Enriques wrote for the English translation of Bonola's book is considerably more forthcoming:

> It seems to me that this account, although concerned with a particular field only, might well serve as a model for a history of science, in respect of its accuracy and its breath of information, and, above all, the sound philosophic spirit that permeates it. The various attempts of successive writers are all duly rated according to their relative importance, and are presented in such a way as to bring out the continuity of the progress of science, and the mode in which he human mind is led through the tangle of partial error to a broader and broader view of truth. This progress does not consist only in the acquisition of fresh knowledge, the prominent place is taken by the clearing up of ideas which it has involved; and it is remarkable with what skill the author of this treatise has elucidated the obscure concepts which have at particular periods of time presented themselves to the eyes of the investigator as obstacles, or causes of confusion. . . . May his devotion stimulate others to pursue with ideals equally lofty the path of historical and philosophical criticism of the principles of science! Such efforts may be regarded as the most fitting introduction to the study of the high problems of philosophy in general, and subsequently of the theory of understanding, in the most genuine and profound signification of the term, following the great tradition which was interrupted by the romantic movement of the nineteenth century. (Bonola 1955, iii–iv)

I think that the references to the eighteenth-century view of progress and the unity of science are particularly significant, because they highlight the fact that those studying the foundation of mathematics had found a way to unify mathematics, something that had been lost in the great expansion of mathematics during the nineteenth century and in the "romantic" attacks on science of the period. Part of this gain in unity, however, comes at the expense of "cleaning up the ideas" of earlier mathematicians, that is, in formulating their ideas in terms of our current concepts and the issues we see as important. Gray's criticism of Bonola's work matches Enriques's assessment very well. Gray says that the main problem with Bonola's account is the tendency of seeing the entire history of the parallel postulate as a single issue, the foundations of geometry (Gray 1987, 57).[1] Thus, Bonola's single-minded emphasis on the foundational issues, which Gray justly criticizes as whiggish, is praised by Enriques for clarifying the issues and providing a unified framework for evaluating the history of geometry.

There is no doubt that Bonola's book has been read as contributing to the foundations of geometry, from Enriques onwards. It is worth pointing out, however, that Bonola "steps out" of purely chronological historical presentation when he introduces a section on "Hypotheses equivalent to Euclid's Postulate," ([1912] 1955, 118–121) in which he introduces a contemporary definition of logical equivalence, and an appendix on "The [logical] Independence of Projective Geometry from Euclid's postulate" ([1912] 1955, 227–237). As Gray notes, the appendix did not appear in the original 1906 Italian edition. Standards of historical writing have obviously changed, but even more important, the purpose for writing a history of mathematics has changed in the last one hundred or so years. If current historians of mathematics are involved at all in debate with philosophers, it is a very different debate from that which occurred at the end of the nineteenth century.

It is also clear that the influential histories of the non-Euclidean geometries were written and published with pedagogy in mind. Here Carslaw, the English translator of Bonola's text, speaks for the mathematical community:

> Recent changes in the teaching of Elementary Geometry in England and America have made it more than ever necessary that those who are engaged in the training of the teachers should be able to tell them something of the growth of that science. (Bonola [1912] 1955, v)

There is also no doubt that the pedagogical literature was influenced by the standard historical account of the development of the non-Euclidean geometries and that his account was influenced by the study of the foundations of mathematics. As I mentioned earlier, the standard undergraduate course in geometry now includes a strong dose of formal axiomatics, but it also includes a very surprising amount of history, far more than is typically found in mathematical textbooks on other topics. Increasingly, formal logic and axiomatics are being introduced independently of geometry in newer textbooks so that these new texts may serve a dual function as an introduction to mathematical logic and to geometry. At the same time, the historical accounts are being left out of these textbooks, so that linkage, which was forged between mathematical logic and geometry by mathematicians writing history at the turn of the century, is now being erased.

THE HISTORY OF GEOMETRY AND OF THE A PRIORI

The standard accounts of the development of geometry connect geometry to logic and to foundational issues in the philosophy of mathematics. I have raised some doubts about these accounts but want to emphasize how influential they were in twentieth-century philosophy of science. Logic was seen as the most important philosophical tool by philosophers of science and by analytic philosophers generally. The historical accounts of the connection between logic and geometry allowed for a reconception of geometry as a formal system, which in turn led to the consideration of geometry as conventional. Back at the beginning of the twentieth century, Henri Poincaré's theory of geometric conventionalism applied the lessons of geometry to the problem of space. Kant's theory of space as necessarily Euclidean had been put into question and a widespread reconsideration of the basic concepts of physics was under way. The development of non-Euclidean geometries and their acceptance led the way for even more revolutionary changes in science. Even when Poincaré's views were not fully accepted, they became the focal point of much discussion, so it is to his views that I now turn.

NOTE

1. Also see Gray ([1979] 1989, 149), where he raises the issue of when the notion of the logical independence of axioms appears.

BIBLIOGRAPHY

Baldwin, James Mark, ed. (1901–1905) 1960. *Dictionary of Philosophy and Psychology.* Vol. 1. Gloucester: Peter Smith.

Bonola, Roberto. (1912) 1955. *Non-Euclidean Geometry: A Critical and Historical Account of its Development.* New York, Dover.

Bourbaki, Nicolas. 1969. *Éléments d'Histoire des Mathématiques.* 2nd ed. Paris: Herman.

Bynum, Terrell Ward. 1993. "The Evolution of Frege's Logicism." In *Logic and Foundations of Mathematics in Frege's Philosophy,* edited by H.D. Sluga. New York: Garland, 193–200.

Carruccio, Ettore. 1964. *Mathematics and Logic in History and in Contemporary Thought.* London: Faber and Faber.

Combebiac, Gaston Charles. 1903. "l' Espace est-il Euclidien?" *L'Enseignement Mathématique* 5: 157–177, 262–278.

Contro, W. S. 1976. "Von Pasch zu Hilbert." *Archive for History of Exact Sciences* 15: 285–295.

Coolidge, Julian Lowell. 1909. *The Elements of Non-Euclidean Geometry.* London: Oxford University Press.

———. (1940) 1963. *A History of Geometrical Methods.* New York: Dover Publications.

Curry, Haskell B. 1963. *Foundations of Mathematical Logic.* New York: McGraw-Hill.

Dauge, Félix. 1896. "Sur la géométrie non-euclidienne." *Mathésis (2)* 6: 7–13.

Freudenthal, Hans. 1957. "Zur Geschichte der Grundlagen der Geometrie." *Nieuw Archief voor Wiskunde* 5: 105–142.

———. 1974. "The Impact of Von Staudt's Foundations of Geometry." in *For Dirk Struik; Scientific, Historical and Political Essays in Honor of Dirk J. Struik.*

edited by R. S. Cohen, J. J. Stachel, and M. W. Wartofsky. Dordrecht: D. Reidel, 189–199.

Graham, Loren, Wolf Lepenies, and Peter Weingart, eds. 1983. *Functions and Uses of Disciplinary History*. Boston: D. Reidel.

Gray, Jeremy J. (1979) 1989. *Ideas of Space: Euclidean, Non-Euclidean and Relativistic*. Oxford: Clarendon Press.

———. 1987. "The Discovery of Non-Euclidean Geometry." In *Studies in the History of Mathematics*, edited by E. R. Phillips. New York: Mathematical Association of America, 37–60.

Greenberg, Marvin Jay. 1980. *Euclidean and Non-Euclidean Geometries: Developments and History*. 2nd ed. San Francisco: W. H. Freeman and Co.

Hoüel, Jules. 1863. *Essai d'une exposition rationelle des principes fondamentaux de la géométrie élémentaire*. Paris: Mallet-Bachelier.

———. 1867. *Essai critique sur les principes fondamentaux de la géométrie élémentaire*. Paris : Gauthier-Villars.

Kline, Morris. 1972. *Mathematical Thought from Ancient to Modern Times*. New York: Oxford University Press.

———. 1980. *Mathematics, the Loss of Certainty*. New York: Oxford University Press.

Manning, Henry Parker. 1901. *Non-Euclidean Geometry*. Boston: Ginn & Co.

Meschkowski, Herbert. 1964. *NonEuclidean Geometry*. Translated by A. Shenitzer. New York, Academic Press.

Nagel, Ernest. (1935) 1979. "Impossible Numbers: A Chapter in the History of Modern Logic." In *Teleology Revisited and Other Essays in the Philosophy and History of Science*. New York: Columbia University Press, 166–194.

———. (1939) 1979. "The Formation of Modern Conceptions of Formal Logic in the Development of Geometry." In *Teleology Revisited and Other Essays in the Philosophy and History of Science*. New York: Columbia University Press, 195–259.

Nowak, Gergory. 1989. "Riemann's Habilitationsvortrag and the Synthetic A Priori Status of Geometry." In *The History of Modern Mathematics, Volume I: Ideas and their Reception*, edited by D. E. Rowe and J. McCleary. Boston: Academic Press, 17–46.

Perry, Earl. 1992. *Geometry: Axiomatic Developments with Problem Solving*. New York: M. Dekker.

Renouvier, Charles. 1889. "La philosophie de la règle et du compas, ou des jugements synthétique à priori dans la géométrie élémentaire." *La Critique Philosophique (n. s.)* 5 (2): 337–348.

———. 1891. "La philosophie de la règle et du compas, théorie logique du jugement dans ses applications aux idées géométrique et à la méthode des géomètres." *L'Année Philosophique* 2: 1–66.

Rosenfeld, Boris Abramovich. 1988. *A History of Non-Euclidean Geometry: Evolution of the Concept of a Geometric Space*. Translated by A. Shenitzer. New York: Springer-Verlag.

Saccheri, Girolamo. 1733. *Euclides Vindicatus*. Chicago: Open Court, 1920.

Sherry, David. 1991. "The Logic of Impossible Quantities." *Studies in History and Philosophy of Science* 22: 37–62.

Sommerville, Duncan M'Laren Young. (1911) 1970. *Bibliography of Non-Euclidean Geometry*. New York: Chelsea.

Stäckel, Paul, and Friedrich Engel. 1895. *Die Theorie der Parallellinien von Euklid bis auf Gauss*. Leipzig: Teubner.

Steiner, Hans-Georg. 1964. "Frege und die Grundlagen der Geometrie II." *Mathematische-Physikalische Semesterberichte* n. s. **11**: 35–47.

Torretti, Roberto. (1978) 1984. *Philosophy of Geometry from Riemann to Poincaré.* 2nd ed. Dordrecht: D. Reidel.

Trudeau, Richard J. 1987. *The Non-Euclidean Revolution.* Boston: Birkhäuser.

Whitehead, Alfred North, and Bertrand Russell. (1910–1913) 1927. *Principia Mathematica.* 2nd ed. Cambridge: Cambridge University Press.

Wilder, Raymond L. 1952. *The Foundations of Mathematics.* New York: John Wiley & Sons.

3   Poincaré's Conventionalisms

Poincaré argues that certain elements of empirical science can be "erected"[1] into principles; that is, they can be taken to be definitely true and never questioned. We might say that they are hardened into principles, or elevated to principles. Arthur Pap used this conventionality of principles in Poincaré to ground his theory of a relativized, or functional, a priori, as will be seen in Chapter 5. The conventional principles stand in for what had formerly been taken to be a priori in that they are taken to be true no matter what happens empirically. These conventional principles can be changed, however, so they are not fixed necessities like the traditional a priori.

While the conventionality of principles seems to fit with the relativized a priori, or, as I prefer, the relativized constitutive elements in science, the main type of conventionalism for which Poincaré is known, geometric conventionalism, has a quite different status and a quite different justification than the conventionality of principles. So it may seem unclear the extent to which Poincaré's views are really compatible with the relativized a priori, or at least it is not clear immediately what kind of relativized a priori he would advocate. Nevertheless, with his geometric conventionalism, Poincaré set out one of the most influential arguments that what Kant took to be fixed and necessary was in fact neither. Here I explore the ways in which Poincaré can be properly said to have a theory of the constitutive elements in science and how his views on these matters fit with his overall philosophy. He advocates a traditional form of the a priori in the limited areas of arithmetic and topology but he seems to reject the a priori in physical theory (e.g., Poincaré holds that we have no a priori intuition of space or time).

Although Poincaré's conventionalism of geometry opens up the possibility of conceptual change by taking away the necessity of Euclidean geometry, he is not attempting to explain conceptual change. Indeed in arguing that Euclidean geometry will (and should) continue to be used in natural science, Poincaré advocates continuity and the cumulative growth of science. The whole motivation for developing a relativized or dynamic theory of the a priori or of constitutive elements seems to be lacking in Poincaré, given that he believed very strongly in the continuity of science and had a cumulative image of science building on past results. He denies the kind of conceptual change that Kuhn made famous and that the theories of the constitutive elements in science is supposed to explain. Poincaré's defense of continuity in science is worth quoting for consideration of its metaphors:

> The advance of science is not comparable to the changes of a city, where old edifices are pitilessly torn down to give place to new, but rather to the continuous evolution of zoologic types which develop ceaselessly and end by becoming unrecognizable to the common sight, but where an expert eye finds always traces of the prior work of the centuries past. It is not necessary to think that the outmoded theories had been sterile and vain. (Poincaré 1913, 208, translation modified)

Superficially, it looks like old theories have been overturned and discarded, but in fact elements of those theories remain in current science. Poincaré clearly rejects the Kuhnian idea of the destructive nature of scientific revolutions. Instead of the destruction of whole neighborhoods to create the boulevards in Paris, we have vestigial elements of former sciences, but much like the human caudal appendage, only experts can see the continuity that is not apparent to the layperson. Three years earlier, in *Science and Hypothesis*, Poincaré's argument for this continuity thesis had been instrumentalist—observations and experimental laws remain intact, even if the theoretical elements change (1913, 140). He even states that science has no business looking for anything beyond what is observable, but in the passage quoted earlier, originally published in 1905, there is no mention of observational versus theoretical elements of science, just a claim that traces remain.

Poincaré's firm commitment to the continuity of science seems to have been developed in response to an argument taken from Tolstoy that spilled out into the popular press in Paris over the bankruptcy of science, an argument made in a famous pamphlet by Ferdinand Brunetière (1895; Poincaré 1913, 362ff., 140).[2] In relation to this debate, Poincaré alludes to what is now called the pessimistic induction (Laudan [1981] 1984), that the past failures of science show that current science is also mistaken. It is in this context that Poincaré says that the past contributions of science have not been "sterile and vain." However, even though Poincaré is not looking to explain revolutionary conceptual change in science, his conventionalism provides an appealing framework for doing so. Fundamental constitutive elements of science can change, leading to profound revolutionary changes in science. What Poincaré shows us is how change can occur at a very deep level that cannot be equated with mere empirical change, since his conventions are isolated from empirical testing. Of course, for Poincaré an even deeper bedrock remains a priori knowledge in the traditional sense in the case of arithmetic (in particular mathematical induction). Therefore, Poincaré has a mixed account, with some traditional a priori elements and some conventions, which can be taken as constitutive elements of scientific theories.

Poincaré's two kinds of conventions play different roles in his philosophy of science. He has a hierarchical picture of the sciences, with arithmetic on the bottom, then geometry, then mechanics, with each level presupposing what comes before (Friedman 1999, 74; Heinzmann and Stump 2014, 4). Clearly then the conventions of metric geometry must come into play prior to the conventions of the principles of mechanics. Both kinds of conventions are open to choice, though this choice is constrained, according to Poincaré, and informed by our experience. Since there are always alternative conventions, there is the possibility that science may have changed conventions over time and this would account for scientific revolutions and for conceptual change. Both with the hierarchy of the sciences and through his principles that are elevated to conventions, Poincaré provides a concrete understanding of what it means for one element of theory to be presupposed by another. In this chapter, I first make my case that there are two distinct kinds of conventions in Poincaré. I then present his argument for geometric conventionalism in two parts, first an argument against the a priori determination of metric and second an argument against the empirical determination of metric. Presentation of Poincaré's two-part argument for the conventionality of metric geometry will show how radically different this type of convention is from the conventionality of principles. Poincaré's views and the misinterpretation of his views had a profound influence on the philosophy of science in the twentieth century.

TWO KINDS OF CONVENTIONS

Poincaré's two kinds of conventions have not been clearly distinguished, neither by Poincaré himself nor by his interpreters.[3] Poincaré argues that some elements of empirical science can be "erected" into principles, that is, they can be taken to be definitely true and never questioned. However, geometric conventionalism has a separate two-part justification that is quite different from his justification of the conventionality of principles. His thoroughgoing relational theory of space and his view that space exhibits only topological properties leads him to the view that the metric geometry of space cannot be determined empirically. This argument against empirical determination of metric is certainly the most important defense of Poincaré's conventionalism, and I discuss it later in this chapter, but Poincaré's argument also includes an a priori argument against Kant's theory of geometry, and this argument should be understood in the context of a philosophical theory of meaning and a formal conception of geometry that was discussed in Chapter 2. Applying an argument about the meaning of terms to Poincaré's argument for the geometrical conventionalism has led to some of the most serious misinterpretations of his view, so I should emphasize here that I completely reject the view that Poincaré's geometric conventionalism is based on any kind of linguistic conventionalism. However, Poincaré does use some arguments concerning the definition of primitive terms in geometry. I put those arguments in their proper context in the explanation of Poincaré's views that follows.

Poincaré quite explicitly distinguishes between the two kinds of conventions as part of the distinction between mathematical theory and physical theory, in the "General Conclusions of the Third Part" section of *Science and Hypothesis*. Not only has it been commonly assumed that there is but one kind of convention in Poincaré and one set of arguments for his conventionalism; it has also frequently been said that Poincaré does not distinguish mathematical theory from physical theory and that if he had done so his conventionalism would be undermined. Missing the distinction between the conventionality of principles and the conventionality of metric geometry has led to many errors of interpretation, so it is important to elaborate the difference. The conventionality of principles is most analogous with the idea that some statements of physical theory function as though they were a priori or are constitutive, although Poincaré does not use this language.

The first thing to notice is that geometry and the principles of physical theory occupy different levels in Poincaré's hierarchy of the sciences. While this in itself does not necessarily imply that they have a different status, it does indicate that they have a different role to play in science. A second important point to notice is that while Poincaré says that principles of physical theory can be taken to be absolutely true, he says that metric geometry is neither true nor false. Furthermore, while conventional principles typically derive from empirical laws, geometry is totally separated from any empirical evidence.[4] The fourth thing to notice is that the argument for the conventionality of metric geometry is quite different from the argument for the conventionality of principles. Rather than hardening an empirical law into a principle, in the case of metric geometry we have a two-part strategy, consisting of an argument, first, against the a priori determination of metric and, second, against empirical determination of metric. Thus, there are major differences between the conventionality of principles and the conventionality of metric geometry and their epistemological status is very different.

I now want to look in detail at the passages where Poincaré makes a distinction between conventional principles and the conventionality of metric geometry. He starts with the distinction between the empirical part of physical theory and the postulates (or principles) of a physical theory:

> The principles of mechanics, therefore, present themselves to us under two different aspects. On the one hand, they are truths founded on experiment and very approximately verified so far as concerns almost isolated systems. On the other hand, they are postulates applicable to the totality of the universe and regarded as rigorously true. (Poincaré 1913, 123–124, translation modified)

Poincaré goes on to describe the postulates as conventions and to explain their role in physical theory:

> If these postulates possess a generality and a certainty which are lacking in the experimental truths from where they are drawn, this is because they reduce in the last analysis to a mere convention which we have the right to make, because we are certain beforehand that no experiment will come to contradict it. (Poincaré 1913, 124, translation modified)

Poincaré also elaborates the nature of conventions and explains the paradoxical idea that principles of mechanics can not only have an empirical origin but can also be immune to empirical refutation. We take them as fixed principles and look elsewhere to deal with negative evidence:

> This convention, however, is not absolutely arbitrary; it does not come from our whim; we adopt it because certain experiments have shown us that it would be convenient. Thus is explained how experiment could construct the principles of mechanics, and yet why it cannot overturn them. (Poincaré 1913, 124, translation modified)

Next—and this is the evidence for the point that I am making here—Poincaré distinguishes the conventionality of principles and the conventionality of geometry. As I noted before, he holds that metric geometry is neither true nor false and that it does not have an empirical origin:

> Compare with geometry: The fundamental propositions of geometry, such as for example Euclid's postulate, are nothing more than conventions, and it is just as unreasonable to inquire whether they are true or false as to ask whether the metric system is true or false. (Poincaré 1913, 124, translation modified)

He goes on to try to preempt any argument that the epistemological status of the principles of mechanics and the principles of geometry are the same. The crucial distinction is that the objects of study of each discipline are quite different:

> At first blush, the analogy is complete; the role of experiment seems the same. One will therefore be tempted to say: Either mechanics must be regarded as an experimental science, and then it must be the same for geometry; or else, on the contrary, geometry is a deductive science, and then one can say as much of mechanics.
>
> Such a conclusion would be illegitimate. The experiments which have led us to adopt as more convenient the fundamental conventions of geometry focus on objects which have nothing in common with those geometry studies; they focus on the properties of solid bodies, on the

rectilinear propagation of light. They are experiments of mechanics, experiments of optics; they cannot in any way be regarded as experiments of geometry. And even the principal reason why our geometry seems convenient to us is that the different parts of our body, our eyes, our limbs, have precisely the properties of solid bodies. On this account, our fundamental experiences are preeminently physiological experiments, which focus, not on space which is the object the geometer must study, but on his body, that is to say, on the instrument he must use for this study.

On the contrary, the fundamental conventions of mechanics, and the experiments which prove to us that they are convenient, bear on exactly the same objects or on analogous objects. The conventional and general principles are the natural and direct generalization of the experimental and particular principles. (Poincaré 1913, 124–125, translation modified)

Thus, Poincaré makes a sharp distinction between the conventionality of principles and the conventionality of metric geometry based on the objects to which they refer. He also notes another distinction between them. While it is right to separate geometry from physical science, it would not be right to separate conventional principles from the empirical theory in which they are found:

Let it not be said that thus I trace artificial boundaries between the sciences; that if I separate by a barrier geometry properly so called from the study of solid bodies, I could just as well erect one between experimental mechanics and the conventional mechanics of the general principles. In fact, who does not see that in separating these two sciences I mutilate them both, and that what will remain of conventional mechanics when it shall be isolated will be only a very small thing and can in no way be compared to that superb body of doctrine called geometry? (Poincaré 1913, 125, translation modified)

This quote implies that the number of principles in physical theory is small and that they cannot stand alone to form an interesting system.

Finally, Poincaré also expresses in a schematic way the process of turning an empirical law into a conventional principle where again we see a difference from the geometric case. Poincaré rejects the idea that geometry comes in any way from empirical findings, while principles precisely do come from empirical findings. Furthermore, there is no residual empirical element in geometry, but there always is such an element in principles. In other words, we cannot turn our entire physical theory into conventions; there is always some empirical content, while in geometry there is none:

How can a law become a principle? It expressed a relation between two real terms A and B. But it was not rigorously true, it was only approximate. We introduce arbitrarily an intermediary term C more or less fictitious, and C is *by definition* that which has with A *exactly* the relation expressed by the law.

Then our law is separated into an absolute and rigorous principle which expresses the relation of A to C and an experimental law, approximate and subject to revision, which expresses the relation of C to B. It is clear that, however far this partition is pushed, some laws will always be left remaining. (Poincaré 1913, 125–126, translation modified)

There are times when Poincaré seems genuinely inconsistent on the truth of what he calls the principles of mechanics. Unfortunately, he sometimes says that the principles of mechanics are neither true nor false because they are conventions, even though he holds that principles are true. In *The Value of Science*, when his discusses principles, using Newton's law of gravity as an example, Poincaré contradicts himself in the space of a few lines:

we may elevate it [a sufficiently confirmed law] into a principle by adopting conventions such that the proposition would be *certainly true* . . . The principle, henceforth crystallized, so to speak, is no longer subject to the test of experiment. It is *not true or false*, it is convenient. (Poincaré 1913, 335, emphasis added and translation modified)[5]

So, while Poincaré is completely clear that the theorems of geometry are conventions and therefore are neither true nor false and that not all of mechanics is conventional but, rather, it is true, he is sometimes ambiguous about the truth of principles. What *should* Poincaré say about the truth of principles? First of all, in opposition to Duhem, he insists that scientific theories are (or can be) true (Poincaré 1913, 323–325). However, Poincaré says that "[t]he truth we are permitted to glimpse is not altogether what most men call by that name" (1913, 206). Poincaré is led to reject metaphysical realism because of his denial of the existence of an "external reality":

No, beyond doubt a reality completely independent of the mind which conceives it, sees or feels it, is an impossibility. A world as exterior as that, even if it existed, would for us be forever inaccessible. (1913, 209; see also 28)

Instead of correspondence, Poincaré emphasizes the unifying power of theory and of what is common to all thinking beings. These views may make it easier for him to claim that principles are true, but is not this also the same for the truth of geometry? Since Poincaré thinks we have clear alternatives in the case of metric geometry; he may have more reason for withholding the label 'true' from it. It is nice for my argument that Poincaré make a distinction between principles, which are true, and geometry, which is neither true nor false, but can he maintain this distinction? It seems to me that this is problematic, given that to the extent that principles are empirical laws, they are true, but to the extent they are conventions, they are neither true nor false.

THE CONVENTIONALITY OF GEOMETRY

As for the arguments for the conventionality of geometry, I have claimed elsewhere that Poincaré has a two stage argument for the conventionality of metric geometry (Stump 1989, 1991, 1996, 1998). First, he argues against any a priori determination of the metric of space, which is one of the many ways that Poincaré directly opposes Kant. He argues that we have no intuition of time and no intuition of space (Poincaré 1899b, 274), completely rejecting Kant's ground for arithmetic and geometry. The existence of consistent non-Euclidean geometries shows us that there is no a priori method of determining the metric of space. Poincaré also has what looks like a formal view of geometry, which may seem incongruous, given his strong antiformalist stand in arithmetic, but nevertheless we can find this in his writings. Second, he also argues against any empirical determination of the metric of space, a point in the argument where Poincaré is often thought to be using something like the Duhem–Quine underdetermination thesis. However, this cannot be the correct interpretation of Poincaré's geometric conventionalism, because all empirical theories, and not just physical geometry, are underdetermined in the Duhemian sense. The major interpretive problem for those who hold this epistemological interpretation of conventionalism is to explain why Poincaré holds that only geometry and a few principles are conventional, not all of science (Stump 1989, 348; Friedman 1999, 73). As we see clearly in Poincaré's critique of Le Roy and even earlier in *Science and Hypothesis*, he firmly rejects the idea that there is a generalized conventionality of science (1913, 321ff., 125, respectively). So we need to find an argument in Poincaré that results in only the metric geometry of space and a few principles being conventional given that underdetermination does not fit the bill since it is a general thesis that applies everywhere. To make matters more confusing, Poincaré clearly does use some form of the underdetermination thesis in the conventionality of principles. It is important to remember though that he still thinks that this argument is quite limited. He seems to say that there are only a small number of conventional principles, and he clearly says that you can never get rid of all the empirical content of your theory by turning empirical laws into principles.

As an alternative, I propose that Poincaré thought that a fully relational theory of space is not only possible, but that in outline it is quite defensible. Thus, in *Science and Hypothesis* Poincaré argues that the concept of absolute space is both "repugnant to the mind" and refuted empirically (1913, 108, also 413). As many authors have noted, a full relational theory of space would allow for conventionality of metric, an argument that I develop in the following. However, it turns out that a fully relational theory of space has not been shown to be true. Therefore, I have argued, Poincaré's argument for the conventionality of the metric geometry of space does not go through. What I have also argued, however, is that Poincaré has in effect been empirically rather than philosophically refuted, which undercuts many of the criticisms that philosophers have aimed at Poincaré through the years—that he is inconsistent, that his theory amounts to trivial semantic conventionalism, and so on.

A PRIORI ARGUMENTS

The existence of consistent non-Euclidean geometries leads Poincaré to the view that the metric geometry of space cannot be determined a priori. Concerning pure geometry, Poincaré holds the modernist sounding views that we have no pre-axiomatic understanding of geometric primitives, that rigor demands that we eliminate appeals to intuition in geometry, and that pure metric geometry is neither true nor false. By calling his view modernist I am referring to the fact that philosophers and historians have recognized that Poincaré expressed a view of geometry that was influential in overturning the traditional view of geometry as a synthetic a priori science of space. Indeed, since Poincaré changes the traditional view of geometry as the a priori science of space so profoundly, Ernest Nagel gives him a leading role in the development of the formal conception of mathematics in his widely cited paper "The Formation of Modern Conceptions of Formal Logic in the Development of Geometry" (Nagel [1939] 1979). Alberto Coffa (1986) extends Nagel's analysis by claiming that there was a crisis in the interpretation of primitive geometrical terms in the nineteenth century and that this crisis led to the development of formal axiomatic systems. References to a central role played by Poincaré can also be found in Joan Richards's (1994) survey and in Jeremy Gray's (2008) account of the development of modernism. Furthermore, despite Federigo Enriques criticism of the view that pure geometry is neither true nor false, he also gives Poincaré a major role in the construction of a modern formal view of geometry (Enriques [1911] 1991).

Along with the wide recognition of Poincaré's influence on a modern conception of geometry, it is well known that Poincaré takes quite a traditional position on arithmetic, holding that the axioms of arithmetic are synthetic a priori truths, that the notion of whole number is irreducible and that we have a special intuitive knowledge of the fundamental principles of arithmetic—mathematical induction and the continuum. The aim of this section is to analyze Poincaré's arguments for the conventional and formal nature of pure metric geometry, and to explain how he can argue for formalism in metric geometry only. We will then be in a better position to understand his conventionalism. While Poincaré was clearly influenced by the development of analytic methods in geometry and the widespread acceptance of non-Euclidean geometries, his use of these developments is original.

THE MEANING OF GEOMETRICAL TERMS

At the turn of the twentieth century there were two parallel debates about the nature of geometry, one between Henri Poincaré and Bertrand Russell and the other between David Hilbert and Gottlob Frege. Poincaré and Hilbert argue for a new conception of geometrical systems, while Frege and Russell defend a traditional way of looking at these systems. The debate takes up the questions, Which geometry is true? and What is the nature of definitions and axioms? To deal with the multiplicity of geometries, questions about their truth, and the meaning of their primitive terms, Poincaré and Hilbert propose a radically new view of geometry. They hold that all that we can say about the meaning of *point, straight line, distance,* and so on is that which we have stated in the axioms of the system and that geometry is not a set of truths about some previously known objects. Thus, Poincaré formulates a new view of geometric theories, that geometry does not express propositions and that there are no special objects that geometry studies. Rather, geometry is just a system of relations that can be applied to many kinds of objects.

Perhaps the best way to see the main point of disagreement between Poincaré and Russell is to examine their exchange on the nature of definitions in geometry. Poincaré challenges Russell to come up with a meaningful definition of the primitive terms of geometry that is independent of the axioms, because he has come to the view that the axioms themselves define the terms. Russell thinks that axioms could not possibly define terms, and that Poincaré is simply presenting an incorrect view of definition (Russell 1899, 699–700). In a sense Russell and Frege are completely correct on this point, since in the current usage of logicians, the primitive terms of a formal system are undefined, not defined by the axioms, but to insist on logical rigor here is to miss the importance of the debate. Since explicit definition cannot account for everything without regress, we are left with a set of primitives that are undefinable. The question of what to do with these undefinable terms is, as Russell realized, exactly the point where he and Poincaré disagree, and it is also exactly the point in which the status of formerly a priori elements of our knowledge is at stake. Russell distinguishes two kinds of definition: philosophical and mathematical. According to Russell, mathematical definition merely tells the relation of one term to another that is already known and that does not give the meaning of the term. However,

> [p]hilosophically, a term is defined when its *meaning* is known, and its *meaning* cannot consist in relations to other terms. One will surely grant that a term cannot be usefully employed unless it signifies something. Its signification may be simple or complex. . . . In the first case, one defines the term philosophically by enumerating its simple elements. But when the term is itself simple, no philosophical definition is possible. (Russell 1899, 700, emphasis in original)

In fact, Russell will argue, no definition of any kind is possible if the term is simple. It should be stressed that Russell never doubts that these terms are meaningful; it is just that the meaning of simple terms is philosophically difficult. An important claim regarding meaning can be seen in the earlier-cited passage. The meaning of all terms must be determined *before* they can be used to express a proposition, a view which Alberto Coffa called the thesis of semantic atomism (1986, 21). While this principle may seem unproblematic, it caused problems in the case of geometry.

Russell holds that geometry must be about some previously determined objects, and that the axioms of geometry express truths. These two points go hand in hand. If the primitive terms are already meaningful, then the axioms must make true or false claims about the objects denoted. Both these principles are central to the traditional view of geometry as a science, and both are accepted by Russell (1899, 702). Because of his acceptance of the traditional view of geometry as a science, that is, as expressing truths, Russell assumes that we must be able to know the meaning of geometric terms, independently of their relations to other terms, as difficult as that may be. The only thing Russell can say at this point is that we have some kind of intuition of geometric objects. Frege argues in the same way in his debate with Hilbert. The problem is that neither Russell nor Frege has much to say about intuition, about what it amounts to, or about why we should depend on it. Direct inspection is supposed to allow us to justify an analysis of a simple term "by the immediate feeling of its correctness" (Russell 1899, 703). Apparently one is to take a suggestion, think about it, and hope for a feeling that it is right or wrong.

Poincaré argues against Russell's view of definition, claiming that the meaning of primitive terms can be fixed only by the axioms of the geometrical system in which we are working. Fixing the meaning of primitive terms by the axioms is the only way to proceed, Poincaré argues, because, if we were to take a primitive term out of the context of an axiomatic system, it would lose all meaning:

> If one wants to isolate a term and exclude its relations with other terms, *nothing* will remain. This term will not only become indefinable, it will become *devoid of meaning*. (Poincaré 1900, 78, emphasis in the original)

Thus, Poincaré holds that outside of the context of an axiomatic system, geometrical primitives mean nothing. The lack of precision that Russell and Frege are willing to accept is totally unacceptable to Poincaré. He also notes rather sarcastically that appeals to intuition put an end to rational argument:

> I find it difficult to respond to those who think they have a direct intuition of the equality of two distances or two durations; we speak very

different languages. I can only envy and admire them, without understanding, since I completely lack this intuition. (Poincaré 1899a, 274)

On the other hand, Poincaré thinks intuition has a role to play in mathematics that cannot be eliminated. While he holds that intuition is unhelpful and even dangerous in geometry, Poincaré thinks that we know a crucial part of arithmetic by intuition—the principle of mathematical induction:

> Why then does this judgment force itself upon us with an irresistible evidence? It is because it is only the affirmation of the power of the mind which knows itself capable of conceiving the indefinite repetition of the same act when once this act is possible. The mind has a direct intuition of this power, and experience can only give occasion for using it and thereby becoming conscious of it. (Poincaré 1913, 39; also see 213, 434)

Poincaré seems as traditional here as Russell and Frege. Poincaré's attitude toward intuition in arithmetic presents striking contrast to his view that intuition of the geometric primitives is completely empty. He thinks that arithmetic is synthetic a priori knowledge and that mathematical induction is the central principle in arithmetic. The issue I want to focus on for a moment is how he came to view geometry and arithmetic so differently.

INTUITION IN MATHEMATICS

Poincaré's basic position on the role of intuition in mathematics is developed in an 1889 article on logic and intuition, and is repeated in several places in his works. While he accepts that intuitive arguments play a role in pedagogy and in mathematical discovery, as well as in the epistemology of arithmetic, Poincaré also recognizes the development in nineteenth-century mathematics toward more rigor, more explicit definitions, more emphasis on proof, and the ferreting out of implicit assumptions. Crucial to this rigor is the idea that we must make all of our implicit assumptions clear. In reviewing Hilbert's *Foundations of Geometry*, Poincaré agrees that we should avoid intuition and depend only on what follows logically from the axioms in geometry:

> Is the list of axioms complete, or have we let escape some that we apply unconsciously? This is what we need to know. To find this out, we have one and only one criterion. We must investigate whether or not the geometry is a logical consequence of the explicitly stated axioms; that is to say, if these axioms, entrusted to a reasoning machine, could produce the entire series of geometric propositions. If they can, we will be certain that we have not forgotten anything, because our machine cannot function except according to the rules of logic by which it was constructed.

It does not know of this vague instinct that we call *intuition*. (Poincaré 1902b, rpt. in Poincaré 1916–1956, XI:96).

The "reasoning machine" Poincaré has in mind is Stanley Jevons's so-called "logical piano," the logic of which is equivalent to Venn diagrams. Jevons demonstrated his machine in 1866 (Jevons 1869, 59–60; [1874] 1958, 170ff.). The machine started out as a "logical abacus," a set of blocks representing subject, predicate, and middle terms. With the addition of levers to move the blocks, Jevons developed a sort of logical adding machine (for a description, see Gardner [1958] 1982, chap. 5.) While the logic of which Poincaré is aware appears to be quite limited, it is important to recognize that Poincaré does speak of a logical element in mathematics and that, broadly speaking, he has a formal conception of geometry.

In his early articles, Poincaré argues that geometry concerns only the relations expressed in the axioms and not some inherent features of the primitives: "What we call geometry is nothing but the study of formal properties of a certain continuous group; so we may say, space is a group" (Poincaré 1898), 41). (We will see later that Poincaré is even willing to give up the notion of a group for something still more general and abstract.) As Poincaré explains in a section of his *Monist* article of 1898 titled "Form and Matter," the idea is that in geometry the properties of the primitives or of the objects to which geometrical relations are applied are not important at all, as far as geometry is concerned. The set of relations that hold between the primitives constitute the form, not the matter, of geometric objects, and these are what is studied:

> The different ways in which a cube can be superposed upon itself, and the different ways in which the roots of a certain equation may be interchanged, constitutes two isomorphic groups. They differ in matter only. The mathematician should regard this difference as superficial, and he should no more distinguish between these two groups than he should between a cube of glass and a cube of metal. (Poincaré, 1898, 40; also see 1897, 62–63)

These views are influenced by Poincaré's work in projective geometry and in group theory. Since projective geometry studies the properties of figures which are invariant under a group of projective transformations, projective geometry itself can be seen as a move away from the classical view of geometry as a science. What is studied in projective geometry is certain invariant relations, not any special objects. The objects under consideration may change, but the relations remain the same (Nagel 1939, 206). It is because of this that Nagel sees the development of projective geometry as part of a move towards formalism, although it is important to note that Klein did not interpret his own project in this way (Klein 1921–23, 241–243, cited in Coffa 1986, 10).

Poincaré's central argument for a formalist view of (pure) geometry is that geometric objects are stipulatively defined from more abstract objects. This is why he can say that space is not real; in the 1889 article he says whole number is the only mathematical concept we need:

> But how has one achieved rigor? By restraining more and more the role of intuition in science, and increasing the role of formal logic. Previously one started from a large number of notions, which were regarded as irreducible and intuitive primitives; such were the notion of whole number, fraction, *grandeur continue*, space, point, line, surface, etc. Today only one remains, that of whole number; all the others are merely combinations, and at this price, we achieve perfect rigor. (Poincaré 1899b, rpt. in Poincaré 1916–1956, XI:129)

There is a strong tradition in nineteenth-century mathematics of interpreting unknown (or seemingly impossible) objects as combinations of known simples. This tradition goes back at least to Hamilton's geometric interpretation of complex numbers and continued in Beltrami and Klein's early treatment of non-Euclidean geometries. Poincaré refers to this tradition in his 1904 article (2–3). In any case, there is no doubt that Beltrami's work influenced Poincaré; consider the section of *Science and Hypothesis* on non-Euclidean geometry where Poincaré introduces Beltrami's model and extends it from two to three dimensions. In the preceding quote, Poincaré takes the position that geometric objects are not real; they stand for complex arithmetical objects. But he also claims that we have an intuition about some abstract (and nonmetric) geometry, first projective geometry and, then, topology. Here is a rather late quote from *Science and Method* (1908) where Poincaré explains again why geometric objects are unreal:

> Here are three truths: (1) The principle of complete induction; (2) Euclid's postulate; (3) the physical law according to which phosphorus melts at 44° (cited by M. Le Roy). These are said to be three disguised definitions: the first, that of the whole number; the second, that of the straight line; the third, that of phosphorus. I grant it for the second; I do not admit it for the other two. I must explain the reason for this apparent inconsistency. First, we have seen that a definition is acceptable only on condition that it implies no contradiction. We have shown likewise that for the first definition this demonstration is impossible; on the other hand, we have just recalled that for the second Hilbert has given a complete proof. As to the third, evidently it implies no contradiction. Does this mean that the definition guarantees, as it should, the existence of the object defined? We are here no longer in the mathematical sciences, but in the physical, and the word existence has no longer the same meaning. It no longer signifies absence of contradiction; it means objective existence. (Poincaré 1913, 468)

Thus, Poincaré accepts a formal view of metric geometry. Metric primitives can be explicitly defined in terms of more basic primitives and a complete proof of the consistency of the metric geometries can be given formally. However, in reply to logicism and formalism in mathematics Poincaré argues that is impossible to prove the consistency of arithmetic without using mathematical induction. Therefore, while one can eliminate metric geometry by taking all of geometry to really concern numbers plus some projective or topological primitives that are nonmetric, one cannot eliminate arithmetic and our intuitive knowledge of mathematical induction. In a second argument presented on the following page, Poincaré repeats his claim that geometric (i.e., metric) terms can be "defined away," while arithmetical ones cannot, arguing that whole number and the principle of mathematical induction have equivalent definitions, but only in virtue of a synthetic a priori judgment, not on the basis of an explicit stipulative definition. On the other hand, in the case of a straight line:

> We have not, as in the preceding case, two equivalent definitions logically irreducible one to the other. We have only one expressible in words. Will it be said there is another definition which we feel without being able to express it in words, since we have the intuition of the straight line or since we represent to ourselves the straight line? First of all, we cannot represent it to ourselves in geometric space, but only in representative space and then we can represent to ourselves just as well the objects which possesses the other properties of the straight line, save that of satisfying Euclid's postulate. These objects are 'the non-Euclidean straights,' which from a certain point of view are not meaningless entities, but circles (true circles of true space) orthogonal to a certain sphere. If, among these objects which are equally representable, we call the first (the Euclidean lines) straight, and not the latter (the non-Euclidean lines), this is surely by definition. (Poincaré 1913, 469–470)

Notice that Poincaré refers back to his debate with Russell over definitions that cannot be expressed in words. Whatever intuition we have about space does not include metric properties. We can represent non-Euclidean as well as Euclidean lines and we have to choose which to call straight. We do not have any such choice in the case of whole number.

After reading Hilbert's *Foundations of geometry*, Poincaré extended his strategy of eliminating primitive terms in geometry. In a review of Hilbert (Poincaré 1902b) and even more explicitly in the nomination that he wrote for the awarding of the third Lobachevskii prize to Hilbert (Poincaré 1904), Poincaré argues that Lie's work contains an artificial limit—space is always considered as a number manifold (*Zahlenmannigfaltigkeit*). Lie is limited to the study of continuous groups but Hilbert shows how to move beyond this limit. However, while the idea of group as a basic primitive concept can be overcome, the intuitive element remains and is simply moved back to a more abstract level, from group theory to topology. Referring to Hilbert's second group of axioms, the axioms of order, Poincaré says,

> The axioms of order are presented as dependent on projective axioms, and they would not have any meaning if one did not allow the latter, since one would not know what three points in a straight line is. And yet, there is a peculiar geometry which is purely qualitative and which is absolutely independent of projective geometry, that does not presuppose as known either the notion of a straight nor that of a plane but only the notions of line and surface; it is what one calls topology. (1904, 8)

How are we to understand Poincaré's arguments that metric geometry is a formal, non-intuitive science, when at the same time he defends intuition in arithmetic. First, Poincaré's arguments for formalism in geometry and against formalism in arithmetic both can seem remarkably question begging. On one hand, his view that we have no geometric intuition seems to be equivalent to his acceptance of the consistency of non-Euclidean geometry. On the other hand, how can Poincaré be so sure that formalization of arithmetic will fail and that the concept of whole number will not change?

Poincaré's argument that the principle of mathematical induction is so fundamental because we must always rely on it is compelling, but given what Poincaré says about the meaning of primitive terms in geometry, I am troubled by the fact that his argument seems to presuppose that we can know in advance what will be called an arithmetical system. Negating the parallel postulate obviously changed the metric properties of geometry, and globally, the geometry discovered by Bolyai and Lobachevskii is very strange. Who could have known that hyperbolic geometry would be accepted as a geometry. It seems compelling to say that we need mathematical induction to show that arithmetic is consistent and that everyone accepts consistency as a requirement. However, the consistency of hyperbolic geometry was only gained by removing metric properties from what counts as geometry. Why could not the same argument be made in arithmetic?

A second problem with Poincaré's argument is how he can maintain a distinction between acceptable and unacceptable forms on intuition. For example, what do we say about the axioms of topology? Or the continuum? We could say (a) that there is no geometric intuition, and that "analytic geometry" is fundamentally arithmetical, rather than geometric. This interpretation fits the early passage that I quoted previously. Or, we could say (b) that there is a limited form of geometric intuition, the analytical or qualitative part. This second interpretation fits the later passages that I have cited and also his remarks on the intuition of the continuum and his rejection of the arithmetization of the continuum (Folina 1992, chap. 6).

Under both interpretations however, Poincaré still has to claim that some intuition is completely bankrupt, while other intuition is a priori knowledge (and thus true and certain) and this seems problematic. Under interpretation (a), Poincaré can at least argue that an entire class of intuitions, geometric ones, are problematic, while another class; arithmetic one, are not. Under (b), the problem is amplified, since Poincaré must say that some geometric intuition results in knowledge claims, while other kinds of geometric intuition is completely bankrupt. I would prefer to find a single neat classification of intuitions that makes the distinction that Poincaré seeks.

The most promising way of understanding Poincaré's rejection of some forms of intuition while maintaining other forms as necessary is to see his view as stemming from a distinction between intellectual and sensual intuition.[6] The neo-Kantians of the end of the nineteenth century had all given up Kant's intuition as a form of sensibility but left a conceptual intuition of the understanding in place and Poincaré seems to be firmly in this tradition:

> The words, point, straight, and plane themselves should not cause any visual representation. They could arbitrarily designate objects of any nature, provided that one can establish a correspondence between these objects such that for all systems of two objects called points there corresponds one, and only one, of the objects called straights. . . . The reasoning ought to be able, according to [Hilbert] to lead to purely mechanical rules, and to do geometry, it is sufficient to apply strictly the rules to the axioms, without knowing what they mean. One will in this way be able to construct all of geometry, I would not exactly say without understanding it at all, since one grasps the logical connection of the propositions, but at least without seeing anything. One could give the axioms to a reasoning machine, for example the logical piano of Stanley Jevons, and one would see all of geometry come out. (Poincaré 1904, 6–7)

Thus, Poincaré completely rejects intuition as a visual representation of geometric objects. However, he claims that we need some conceptual intuition in order to understand mathematics, even if we need only understand what is explicitly in the axioms. We can credit Poincaré with having clear categories acceptable and unacceptable intuition, although I do not understand how Jevons's logical piano could have the necessary a priori intuition of topology and arithmetic. Presumably, Poincaré is simply trying to explain the formal aspect of metric geometry in this passage (which he published twice).

A third and final problem is that Poincaré's argument in favor of a priori intuition is mostly negative; that is, he is only able to tell us that such an intuition is necessary to understand mathematics. Any attempt by Poincaré to argue that the mind has special intuitive capacities must be suspect, because he definitely changed his view on what is taken as primitive and intuitive. Originally, geometry is nothing but a group, but later it is even less than that! Does not Poincaré's change of mind on groups give one more example of the bankruptcy of intuition and show that he really does not know what he claims to know intuitively? There is little doubt groups were taken to be special intuitive objects in Poincaré's earlier writings. Here is Picard's account of his view:

> Others attribute a lesser or greater role to the mind thinking about the objects of experience. For some of them even, like M. Poincaré, the concept of group, to which we will return in a moment, preexists in our mind and imposes itself as a form of our understanding; moreover, several interpretations of experience are possible, and among these the mind chooses the most convenient and simple.[7]

There can also be no doubt that Poincaré gave up the notion of group as a primitive in geometry after reading Hilbert and took topology as basic. Will he now say that it is a topological concept that preexists in the mind? Indeed, Poincaré even says that some topological properties of space are conventional, as well as intuitive and necessary, in particular, the number of dimension of space is conventional on his view. However, he may be excused for such inconsistent views in this case, since Poincaré probably did not know that dimension number is topologically invariant. The first proof of that fact was given by Brouwer, in Dutch in 1911, one year before Poincaré's death (see Johnson [1979] 1981).

In pure geometry as well as in physical science, Poincaré expresses a formal conception of theories, which he sees as axiomatic systems with arbitrary starting points. The intuition necessary to begin such a formal system is arithmetic and topological, so arithmetic maintains a necessary and nonconventional character. Since he holds a formal conception of metric geometry and a traditional view of arithmetic, we can see that Poincaré is an important transitional figure in the formalization of mathematics, just as he was in the development of the theory of relativity. The crucial point is the connection between a formal conception of science and the treatment of what had been necessary and certain a priori knowledge. Kant's theory of geometry as the a priori theory of space, which is given in intuition, is totally overturned in Poincaré's geometric conventionalism. This leads the way for a functional theory of a priori knowledge, but first we must consider the arguments that Poincaré makes against the empirical determination of metric.

POINCARÉ'S ARGUMENTS AGAINST AN EMPIRICAL DETERMINATION OF METRIC

We have just seen Poincaré's a priori arguments for the existence of non-Euclidean geometries. Poincaré rejects the idea that we have any intuitive understanding of the primitive terms of metric geometry, arguing instead that all we can say about these terms is whatever is given in a system of axioms. To hold that Euclidean and non-Euclidean geometries are interchangeable conventions, Poincaré needs to claim the metric of space is not objectively determinable; that is, that statements about metric are neither true nor false.

Many arguments for the conventionality of metric that have been attributed to Poincaré are inadequate to establish this thesis. For example, Morris Schlick (1935) claims that Poincaré's argument depends essentially on the view that a set of axioms implicitly define the primitive terms in the axioms. Louis Rougier (1920, 124ff.) and Ernest Nagel (1979, 262) also claim that this argument plays an important role. I claim that while this argument plays a role in Poincaré's argument against the traditional view that geometry aims at discovering truths about special objects, he has further arguments for his conventionalist thesis. Indeed, interpreting conventions as definitions is not enough to establish that these statements are neither true nor false. They could be held as true by definition, or simply as 'hard-core' hypotheses, if we the view that there are hypotheses in science that, while still empirical hypotheses, are not called into question at a given moment of testing, so that other (peripheral) hypotheses may be tested as in Quine or in Lakatos ([1970] 1982). Lakatos holds that research programs, not individual hypotheses, are tested, but my point in making the comparison is that principles of mechanics would, on the face of it, be true or false even if we remove them from consideration in empirical testing.

On the other hand, Duhemian type underdetermination arguments go too far to establish Poincaré's thesis. Duhemian arguments would establish that metric is not objectively determinable, if they are interpreted as showing that any metric may be held, as long as one is willing to change enough auxiliary theory.[8] However, as noted earlier, if we interpret Poincaré as using such an argument to defend his thesis of the conventionality of metric, it becomes difficult to understand how he could use such arguments only in the case of metric. Since Poincaré does not deny the truth of physics in general, conventionality of metric cannot simply be a special case of a global conventionalism, as has often been thought to be the case.[9] Instead, Poincaré has specific arguments for the claim that the metric of space is not determinable. Relationalism provides the key to understanding Poincaré's empirical arguments for conventionalism. He does not understand applied geometry as we do now, that is, as the geometry of physical space. Moreover, Poincaré goes even further than current Einsteinian theories do in rejecting Newtonian absolute space. He holds a purely relational view of space, that there is no physical space and that all of mechanics should be described in terms of the relations of physical objects.[10] These relations between physical objects are expressed in geometric terms, but, according to Poincaré, any geometric properties expressed in these relations are artifacts of our description.

In clarifying Poincaré's position on space we can answer charges of inconsistency. For example, Stanley Goldberg notices that Poincaré is a realist about physics and a conventionalist about geometry, a position that Gold-berg calls "wavering between conventionalism and realism" (Goldberg 1967, 938, 939; Goldberg 1970, 76–77, also 75). However, Poincaré does not waver; he is consistently realist about what he calls "real generaliza-tions" in physics, and consistently conventionalist about geometry and a few fundamental principles of physics. In a similar vein, Lawrence Sklar charges that Poincaré wavered between reductionism and conventionalism (1974, 128). Poincaré does hold a reductionist view regarding some theo-retical entities. He rejects mechanistic philosophy of science and holds that some mechanical models are strictly equivalent. The term *mechanical models* (or *physical models*) in discussion of nineteenth century physics refers to the British tradition (Faraday, Maxwell) of making a representation in a physical sense that provides a mechanical explanation of unobserved pro-cesses. Poincaré calls mechanical models "indifferent hypotheses," because it is sometimes possible to change the mechanical model associated with a theory while the quantitative laws remain the same. For example, the wave and particle theories of light result in the same laws of propagation as was proven by W. R. Hamilton in 1832 (Giedymin 1982, 53ff.). Poin-caré distinguishes three kinds of hypotheses: "natural," "indifferent," and "real generalizations." In the case of geometry, Poincaré holds that we must choose between geometries that are really different. Given his relationalist premises, this position is consistent.

THE NATURE OF SPACE

Poincaré holds the nominalist sounding view that we "impose" space on nature (Poincaré 1913, 29, cf. 337). In his preface to the English edition of *Science and Hypothesis*, Poincaré expresses his position rather brusquely: "Space is only a word that we have believed to be a thing" (Poincaré 1913, 5). Mach, who also holds that there is no substantive space, thinks that space is nothing over and above the relations of physical objects (Mach [1893] 1942). In this purely relational sense, Poincaré does think that space exists. The issue here is whether there is anything above and beyond the relations of physical objects such as an absolute space or a physical space that affects physical objects and that could be studied as physical geometry. Stated another way, the issue is how much of the mathematical structure used in physics is to be interpreted realistically, an issue that is of particular inter-est to Poincaré. Michael Friedman expresses the issue of whether the math-ematical structure corresponds to the real structure of the universe this way:

> an infinite number of other quantities and relations can be defined in R^4 . . ., and not all of these represent physically real properties and relations. Thus, from a modern point of view, . . . sameness of spatial position represents no physically real relation, although from Newton's

original point of view it does: it is just the relation of being at rest relative to absolute space.

... physicist themselves distinguish between aspects of mathematical structure that are meant to be taken literally—that really correspond to pieces of the physical world—and aspects of mathematical structure that are purely representative. What is the rationale for this distinction? (Friedman 1983, 337)

Friedman goes on to give a general answer to this question: that the unifying power of theoretical structure shows that it is to be interpreted literally. I question whether there are such general criteria for the existence of theoretical entities, but I do not debate the issue here. My point is that science itself sometimes claims that elements of the mathematical structure of theories are not to be taken literally, leaving open the possibility of an interpretation of science that includes functionally a priori principles. Like Friedman, Poincaré pays close attention to the results of science in making the decision as to which aspects of theoretical structure are to be taken literally.

A RELATIONALIST PROGRAM FOR PHYSICS

In *Science and Hypothesis*, Poincaré argues for pure relationalism, claiming that all of mechanics can be expressed in terms of the relations of physical objects. The key issue in the debate between relational and substantial interpretations of space is the status of inertial effects. Poincaré falsely claims that

[t]he acceleration of a body depends only upon the position of this body and of neighboring bodies and upon their velocities. Mathematicians would say the movements of all the material molecules of the universe depend on differential equations of the second order. (Poincaré 1913, 94)

Poincaré's idea seems to be that the Galilean relativity of space and motion (i.e., velocity) implies the relativity of acceleration, because acceleration is defined (mathematically) in terms of velocity. Here we see that Poincaré, like all the participants in the traditional debate, is misled by his three-dimensional point of view, in which velocity and acceleration must either all be relative or all absolute. Only a four-dimensional point of view allows for the possibility of relativized position and velocity while acceleration remains absolute (Friedman 1983, 228, 16–17). Poincaré calls his law of relativity a generalized law of inertia and claims that in astronomy, his law is empirically confirmed (Poincaré 1913, 94–96).[11] It is well known, however, that acceleration cannot be relativized without the introduction of universal forces (Sklar 1974, chap. 3; Friedman 1983, 21, 120, 297ff.).

Mach does incorporate universal forces in his defense of relationalism. He argues that we cannot know whether it is absolute space or distant objects plus a universal force that account for inertial effects (Sklar 1974, 201). Developed in a Machian way, relationalism becomes an epistemological issue, connected to Duhem's thesis of the underdetermination of empirical theories. Schlick is responsible for interpreting Poincaré's geometric conventionalism in this way and his interpretation became the standard used by the Logical Empiricists and their followers (Friedman 1999, 72).[12] As will be seen in Chapter 5, Arthur Pap (1946) follows a similar line of thought in discussing the role of principles in physical theories, and for principles, there is something correct about this interpretation of Poincaré. However, it cannot be the correct interpretation of Poincaré's geometric conventionalism, as I noted earlier, because all empirical theories, and not just physical geometry, are underdetermined in the Duhemian sense, and Poincaré holds that only metric geometry is conventional, plus a few principles.

UNDERDETERMINATION ARGUMENTS

Poincaré does present two arguments for the conventionality of metric that have been interpreted as defending conventionalism by means of the introduction of ad hoc hypotheses. In his famous parable of a non-Euclidean world, he introduces temperature as a "distorting influence." The parable, however, is part of an argument to prove that a non-Euclidean world is imaginable:

> Nothing then prevents us from *imagining* a series of representations similar in all points to our ordinary representations, but succeeding one another according to laws different from those to which we are accustomed. . . .
> Suppose, for example, a world enclosed in a great sphere and subject to the following laws: The temperature is not uniform. (Poincaré 1913, 75, emphasis added)

The background here is Kant's view that space is a form of intuition such that all of our representations must appear in Euclidean space. In opposition to the Kantians, Poincaré intends to show that it is possible to imagine a non-Euclidean world. The conventionalists must refute a priori arguments that space is Euclidean if they are to uphold the view that space is conventional, so this argument does have an important role to play in the debate over conventionalism. Poincaré rejects the view that the axioms of metric geometry constitute a synthetic a priori in either the classical rationalist or the Kantian sense. Indeed, his introduction of a model to prove the consistency of hyperbolic geometry shows that we cannot use a priori arguments to decide which metric geometry is true (Poincaré 1913, 59–60).[13] However, these arguments does not show that Poincaré would allow the use of universal forces in physics, since the arguments only plays a role in the debate over what kinds of geometry we can imagine and perceive.[14]

Another of Poincaré's arguments that has been interpreted as introducing ad hoc hypotheses into physics concerns the definitions of primitive terms in physics. Lobachevskii proposed using astronomical measurements to decide between Euclidean and non-Euclidean geometry. In a passage that has been the subject of much controversy, Poincaré argues that the most such measurements can tell us is what ratios obtain between certain physical objects, in this case light rays:

> If Lobachevskii's geometry is true, the parallax of a very distant star will be finite. If Riemann's is true, it will be negative. These are the results which seem within the reach of experiment, and it is hoped that astronomical observation may enable us to decide between the two geometries. But what we call a straight line in astronomy is simply the path of a ray of light. If, therefore, we were to discover negative parallaxes, or to prove that all parallaxes are higher than a certain limit, we should have a choice between two conclusions: we could give up Euclidean geometry, or modify the laws of optics, and suppose that light is not rigorously propagated in a straight line. (Poincaré 1913, 81)

Adolf Grünbaum argues against the usual interpretation that this passage shows that Poincaré depends on an underdetermination type argument to support his conventionalism, claiming instead that conventionality of metric follows directly from consideration of the continuity of space:

> Poincaré's much-cited and often misunderstood statement concerning the possibility of always giving a Euclidean description of any results of stellar parallax measurements is a less lucid statement of exactly the same point made by him with magisterial clarity in the following passage: 'In space we know rectilinear triangles the sum of whose angels is equal to two right angles; but equally we know curvilinear triangles the sum of whose angles is less than two right angles. . . . To give the name of straights to the sides of the first is to adopt Euclidean geometry; to give the name of straights to the sides of the latter is to adopt non-Euclidean geometry. So that to ask what geometry it is proper to adopt is to ask, to what line is it proper to give the name straight? It is evident that experiment can not settle such a questions.' (Grünbaum 1968, 106; Poincaré 1913, 235)

Even Poincaré's claim that we could "modify the laws of optics" might be interpretable as nonempirical if the changes he has in mind are changes in the definitions of terms that could themselves be interpreted as nonempirical. Poincaré's argument above amounts to a claim that acceleration can be defined in a non-empirical way in Newtonian theory, a position for which he does make an extended argument (Poincaré 1913, 94ff.). However, it is not at all evident that experiment cannot determine the metric of space, and

Poincaré's position is difficult to maintain according to current theory, given that it depends on a relationalist theory of space.

CONVENTIONALISM AND RELATIONALISM

Friedman sums up the problem with Grünbaum's argument for "metric amorphousness" as follows:

> If we are to draw any conclusions about "metrical amorphousness" from the continuity of physical space, we need an additional premise: namely, that the only spatio-temporal relations that objectively exist are topological relations and order relations. It follows from this claim that properties and relations not definable in terms of topological properties and order relations do not objectively exist and, consequently, that a continuous space objectively lacks metrical properties. Thus, Grünbaum's conventionalism is actually a species of what we have called ideological relationalism. He holds that there is a privileged set of basic spatio-temporal properties and relations must be definable from or reducible to these privileged relations. (Friedman 1983, 304)

Poincaré seems more of an ontological relationalist; that is, denying the existence of physical space and claiming that all of mechanics should be (and can be) described in terms of the relations of physical objects (or occupied spacetime points; see Friedman 1983, 217, 221). Relationalism is indeed a very powerful premise in the debate over the conventionality of metric and, as we saw above, Poincaré thinks that relationalism is empirically confirmed. Indeed, Poincaré's argument that acceleration can be defined in a nonempirical way in Newtonian theory depends entirely on relationalism.

The burden of proof is thus to show that relationalism holds, allowing an argument for conventionality of metric which is independent of underdetermination type arguments. I sketch Arthur Fine's interpretation of how Grünbaum's argument for conventionalism depends on relationalism and then apply the same reasoning to Poincaré's argument for conventionalism.

> At this juncture GC [geochronometric conventionalism—Grünbaum's view] takes two important steps. First, it says that the very existence of metric or congruence relations between spatial intervals depends on the relations that such intervals bear to an extrinsic metric standard which is applied to them. Thus GC steps into the relational parlor. Second, GC asserts that the choice of a particular external standard from among the available alternatives is a matter of convention. (Fine 1971, 462)

Poincaré's argument can be reconstructed along the same lines as Grünbaum's. He argues that magnitude can only be defined as a comparison of two physical bodies, thus arriving at the first step enunciated by Fine earlier (Poincaré 1913, 49–51). This move is clearly relational, because it allows only the relations of physical bodies (or, the relations of occupied spacetime points) to be quantified.

Regarding the second point, Poincaré argues that we have no intuition of distance, and that therefore we have no a priori method of choosing an external standard to define distance (Poincaré 1899a, 274; Poincaré 1900, 77; Poincaré 1913, 416–418). Furthermore, Poincaré argues that given a relationalist theory of space there is no spatial feature that determines an external standard and, therefore, alternative metrics are possible. Anyone holding a substantivalist theory of space would counter that a relationalist theory of space cannot account for all spatial features and that adding enough structure to account for all spatial features also provides enough structure to determine a congruence standard and a metric (Fine 1971, 465). It is important to notice that in *Science and Hypothesis*, Poincaré's defense of the conventionality of metric moves immediately to a discussion of the relativity of space (Poincaré 1913, 83). However, as we saw earlier, Poincaré wrongly thinks that he can account for all spatial features in a purely relationalist framework. Given this premise, he does not need underdetermination arguments. Of course, the burden of proof for his argument falls entirely on relationalism. Friedman analyzes the burden of proof as follows:

> Why should we believe that our primitive or basic spatio-temporal properties and relations include only topological and ordinal ("causal") relations? . . . From our present point of view, these are just properties and relations postulated by our most well-confirmed space-time theories: properties and relations of maximal explanatory or unifying power. Hence, from the present point of view, we can question the objectivity of a given element of theoretical ideology in only one of two ways: we can argue on internal grounds that the theory in which it is embedded is strongly unverifiable (this is what happens to our paradigmatic indeterminate entities from the development of relativity); or we can find a better theory that avoids the question altogether (this is what happens to absolute simultaneity, for example). (Friedman 1983, 305)

Friedman defines "strongly unverifiable" as follows:

> Assertions about absolute velocity or gravitational acceleration are *strongly* unverifiable: they generate no empirical consequences even in conjunction with the rest of relevant theory. Assertions about the geometry of physical space, on the other hand, are only *weakly* unverifiable: to be sure, they generate no empirical consequences in isolation, but they do generate such consequences in conjunction with the rest of relevant theory." (Friedman 1983, 299, emphasis in the original)

Poincaré chooses only the first option. He thinks that relationalism is empirically confirmed and that even acceleration and rotation can be treated as relative motions. Unfortunately, both the Special Theory of Relativity and the General Theory of Relativity (our best spacetime theories) fail to be relational in a sense that will support the conventionality of metric. In the Special Theory of Relativity spacetime is a four-dimensional Euclidean manifold that clearly has a determinate metric. In the General Theory of Relativity, spacetime also has a definite, although variable and non-Euclidean, metric (Friedman 1983, 16–27). I defend Poincaré against the charge of global conventionalism while admitting that his arguments are inadequate to establish a relational theory of space and, therefore, inadequate to establish conventionality of metric.

POINCARÉ'S REJECTION OF AD HOC ARGUMENTS

It is important to point out that there are several ways in which the use of underdetermination-type arguments contradict Poincaré's general views. First, as noted before, the fact that Poincaré rejects Le Roy's total nominalism shows that he cannot use underdetermination arguments to argue for his conventionalism. Second, Poincaré does accept empirical evidence as decisive. He goes so far as to say that "[e]xperiment [L'expérience] is the sole source of truth" (Poincaré 1913, 127), and he is often willing to change his mind on the basis of empirical evidence. For example, after the work of Jean Perrin, Poincaré accepted the existence of atoms, and we will see another case, which follows, where Poincaré considers relationalism to be in question as a result of experimental results. It is also worth noting that in at least one case, Poincaré accepts the refutation of a fundamental principle of physics, what he calls a "natural hypothesis," by empirical theory. When he accepts the quantum theory as presented in the first Solvay conference in 1911, Poincaré gives up the hypothesis of continuity (McCormmach 1967, esp. 43, 54–55).

Evidence against the epistemological interpretation of Poincaré's conventionalism may also be found in the fact that Poincaré criticizes Lorentz for introducing ad hoc hypotheses. Referring to the successive negative results of experiments designed to measure interaction with the ether, Poincaré charges that

> a general explanation has been sought, and Lorentz has found it; he has shown that the terms of the first order must destroy each other, but not those of the second. Then more precise experiments were made; they were also negative; neither could this be the effect of chance; an explanation was necessary; it was found; they are always found; hypotheses are the resources which are lacking least. (Poincaré 1913, 147, translation modified)

Poincaré is objecting to Lorentz's 1895 theory; we will see that he accepts Lorentz's "second order" theory of 1904. For more on Lorentz, see (Goldberg 1969, esp. 987).

There is something artificial in the process of framing a new hypothesis after each negative experimental result. Lorentz takes heed of this criticism and attempts to derive his explanation of the negative result from more general laws in order to show that it is not ad hoc (Lorentz [1904] 1952, 13; 1921). It is implausible that Poincaré would use a general argument concerning the underdetermination of empirical theories to defend his conventionalism, as the epistemological interpretation of conventionalism implies, while criticizing Lorentz for introducing ad hoc hypotheses.

As we saw earlier, in *Science and Hypothesis* Poincaré's argument for relationalism does not involve universal forces, and he wrongly thinks that a pure relationalism in empirically confirmed. We see how central relationalism is to his position when we consider that what Poincaré calls the "principle of relativity" is relationalism, something that is clear from the fact that he returns again and again to the problem of inertial effects in defending his principle and from some of his explicit statements of it:

> The laws of the phenomena which will happen in this system will depend on the state of these bodies and their mutual distances; but, because of the relativity and passivity of space, they will not depend on the absolute position and orientation of the system.
>
> In other words, the state of the bodies and their mutual distances at any instant will depend solely on the state of these same bodies and on their mutual distances at the initial instant, but will not at all depend on the absolute initial position of the system or on its absolute initial orientation. This is what for brevity I shall call the law of relativity. (Poincaré 1913, 83, emphasis removed)

Remember that Poincaré thinks that velocity and acceleration must either all be relative or all absolute, so that rejection of absolute space amounts to a rejection of any physical space. 'Distance' appears here solely because Poincaré has adopted the fiction of a Euclidean metric. He goes on to argue that the same holds from the non-Euclidean standpoint and, finally, in the passages which I criticized earlier, to argue that he can relativize rotation and acceleration (Poincaré 1913, 94). In many places in his works, Poincaré repeats his claim that relationalism is empirically confirmed. Indeed, Goldberg claims that he never retreats from the view that his principle of relativity is empirically confirmed (Goldberg 1969, 937n29). Following Holton, Goldberg uses this fact to distinguish Poincaré from Einstein on relativity, pointing out that Einstein took his own principle of relativity as a postulate. Holton says of Poincaré that

> he made a very significant confession. The news reached him that the results obtained by the great experimentalist W. Kaufmann (1906)

disproved the relativity theory of Lorentz (as well as that of Einstein). To Poincaré, the principle of relativity, on which he had based his own great work published that same year in *Rendiconti,* was immediately suspect, and he now wrote: "Then the principle of relativity would not have the rigorous value we were tempted to attribute to it."[15]

Einstein, on his side, did not take any public notice of Kaufmann. With the characteristic certainty of a man for whom the fundamental hypothesis is *not* contingent either on experiment or on heuristic (conventionalistic) choice, Einstein waited for other to show, over the next years, that Kaufmann's experiments had not been decisive. (Holton 1964, 264)

Holton makes no mention of the fact that Poincaré's next sentence reads "However, before definitely adopting this conclusion, a little reflection is necessary" and that Poincaré goes on to politely question Kaufmann's results and to hope that the experiment can be repeated by other researchers. If Poincaré does indeed use *relativity* to mean pure relationalism, he is right to suspect that it might be empirically disconfirmed, for it has been, according to both the Special and the General Theory of Relativity. Indeed, it is odd to hear Poincaré criticized both for taking experimental evidence too seriously, as well as for not taking it seriously enough!

Only once, to my knowledge, did Poincaré change his position on the status of relationalism. In an essay written in 1912, the last year of his life, Poincaré reviewed his position on space and time in the light of current research, especially that of Lorentz, and reconsidered the problem of inertial effects. He still holds that inertial effects can be defined away mathematically if we consider an isolated system: "since we consider our system to be completely isolated in space, considering it as the entire universe, we have no means of knowing whether it rotates," and he makes no attempt to account for the observable effects which acceleration and rotation produce, apparently still holding that they can be defined away mathematically (Poincaré 1963, 21). Poincaré admits that the Lorentz-Fitzgerald contraction must be assumed when an isolated system is not being considered (Poincaré 1963, 23). At this point, Poincaré seems to consider that the contraction is confirmed and that there is a well-defined theory in which bodies really do contract, so that the Lorentz-Fitzgerald contraction is no longer an ad hoc hypothesis.[16] However, this admission would amount to a rejection of conventionalism. The conventionalist thesis demands that laws of motion be preserved by all arbitrary continuous transformations, not only the Lorentz transformations (Friedman 1983, 17, 7).[17] In any case, this one essay can hardly be used as evidence that Poincaré is a "global conventionalist," when we consider that his arguments for relationalism and conventionalism were developed well before 1912, and that the later position he adopted may not be conventional at all. Whether Poincaré would continue to hold the relationalist position in the light of more empirical evidence and the working out of the limits of that position is a matter of speculation.

POINCARÉ'S PRAGMATISM

Despite the apparently contradictory nature of Poincaré's claims, I think his views exhibit far more consistency than he is often credited. Poincaré is very pragmatic about what will be considered primitive—the absolute starting points—but he always insists that there is a core of such primitives that cannot be eliminated. If Hilbert has shown a way of eliminating the concept of a group, he has merely pushed the primitive concepts required for geometry back to an even more abstract level and this clarification of primitive concepts is to be welcomed. The ability to change the primitives is part of Poincaré's formal conception of geometry, and this view is consistent with his conception of science in general. In particular, Poincaré's pragmatic attitude toward the primitives of geometry is consistent with his other discussions of primitive concepts and basic principles. In mathematical settings as in physical ones, Poincaré is really very open-minded. As we saw above, Poincaré accepts the refutation of a fundamental principle of physics, what he calls a "natural hypothesis," by empirical theory when he accepts the quantum theory as presented in the first Solvay conference in 1911, giving up the hypothesis of continuity. Thus, Poincaré's attitude about his primitives and the basic intuitive starting points in science is not dogmatic.

CONCLUSION

Given that Poincaré's arguments for relationalism fail, it may seem that the only way to save the conventionality of metric geometry is to resort to the global underdetermination argument mentioned earlier.[18] While the epistemological interpretation of conventionalism provides a strong defense, it is not compatible with Poincaré's general views. The misrepresentation of Poincaré's philosophical views has made him look inconsistent, and inconsistency is arguably worse than holding a view that has been disconfirmed by empirical theory, which is all that is implied by interpreting Poincaré as basing his conventionalism on a relational theory of space.

We are left with the idea of metric geometry as a formal system, one in which the choice of axioms determines not only which theorems are true but also the meaning of the primitive terms of the system. Geometric conventionalism is not vindicated, but it is a tantalizing view, given that it could be true if a fully relationalist view of space were possible. However, the idea of geometry being constitutive of physical theories of space (or spacetime) remains. Geometry functions as a priori knowledge but has changed historically from Euclidean geometry to Riemannian, and even if the ultimate reason for the change can in some sense be called empirical, geometry still plays a special epistemological role. Furthermore, the conventionality of principles is still an open possibility, even if Poincaré's geometric conventionalism is unsustainable. Poincaré's philosophy of science can still rightly be seen as inspiring those who would like to understand scientific revolutions by considering the changes in what had been taken to be a priori knowledge. In Chapter 5 we will see how some of Poincaré's ideas were developed into a pragmatic or functional theory of the a priori (really a pragmatic theory of the constitutive elements in scientific theories), with Arthur Pap making the most explicit reference to Poincaré. First, however, I consider the views of those who opposed the Kantian idea of a priori knowledge.

NOTES

1. The French word is *érigées*, from the verb *ériger*, which could mean to erect a monument, to give something a higher status, or to establish an institution. All translations from original sources are by the author.
2. I thank an anonymous referee for Routledge for the Brunetière citation; also see Harry Paul (1968).
3. Jerzy Giedymin does distinguish between the conventionality of metric geometry and the conventionality of principles in Poincaré, but in calling the latter generalized conventionalism he is misleading, since Poincaré thinks that all conventions are limited (Giedymin 1977, 273; 1982).
4. Milena Ivanova (2015) takes this point a step further, insightfully noting that while the source of conventional principles is empirical, the source of geometry is a priori, from the concept of a group.
5. The French is worth considering here, since I am accusing Poincaré of inconsistency: "on peut l'ériger en *principe*, en adoptant des conventions telles que la proposition soit certainement vraie . . . Le principe, désormais cristallisé pour ainsi dire, n'est plus soumis au contrôl de l'expérience. Il n'est pas vrai ou faux, il est commode" (Poincaré 1905, 165–166). The original text is from *Revue de Mètaphysique et Morale*, and it is identical (Poincaré [1902a] 1913).
6. Michael Detlefsen suggested such a distinction in his talk at the Poincaré congress in 1994.
7. "D'autres attribuent un rôle plus ou moins grand à l'esprit travaillant sur les données de l'expérience. Pour certains même, comme M. Poincaré, le concept de groupe, sur lequel nous reviendrons dans un moment, préexiste dans notre esprit et s'impose comme forme de notre entendement; en outre, plusieurs interprétations de l'expérience sont possibles, et parmi celles-ci l'esprit a choisi la plus commode et la plus simple" (Picard 1901, 3).
8. Duhem himself thought that "good sense is the final judge of which hypotheses ought to be abandoned," and he criticizes Poincaré for holding that conventions are not refutable ([1905] 1954, 212–216).
9. As noted earlier, see Poincaré (1913, 321ff., 125), Stump (1989, 348) and Friedman (1999, 73). Giedymin seems to understand that Poincaré's rejection of Le Roy shows that he must not use an underdetermination argument (Giedymin 1977, 283; 1982, 18), but he nevertheless argues in favor of an epistemological reading of Poincaré's conventionalism (Giedymin 1977, 292–293; 1991, 1, 7).
10. As far as I can determine, Arthur Fine (1971) was the first to make explicit the dependence of metric conventionalism on relationalism. Also see the references in Friedman (1983, 304).
11. In other passages, Poincaré claims that inertial effects could be accounted for by postulating some previously undiscovered bodies, thus coming close to

a Machian thesis (Poincaré 1913, 97). However, he also claims that this move is not necessary (Poincaré 1913, 114). Compare (Poincaré 1963, 21, 23) for the one place where Poincaré may change his mind and see DiSalle (2006, chap. 3) for a further discussion of these issues.

12. Friedman discusses the epistemological argument especially with regard to Reichenbach, but considers this argument as a standard line used by everyone including Poincaré (Friedman 1983, 296–301). Indeed, almost everyone except Adolf Grünbaum interprets Poincaré as using the epistemological argument for conventionalism, see Massey (1964, chap. 3) for a survey of the literature and Grünbaum (1968, 106ff.) for his position.

13. See Tuller (1967, 165ff.) and Torretti ([1978] 1984, 136–137) for descriptions of Poincaré's model; see Milnor (1982) for the relations of models introduced by various mathematicians.

14. Grünbaum makes a similar point but argues that Reichenbach's parable argument is also metaphorical (1973, 82).

15. Holton (1964, 264), Holton's emphasis. Holton quotes the original French, H. Poincaré, *Science et Methode* (Paris: Flammarion, 1908, 1920), p. 248. I quote the English translation (Poincaré 1913, 507).

16. Lorentz's position is that the contractions are a physical effect to be explained in terms of the interaction of molecules. By contrast, in Einstein's Special Relativity Theory the contraction is an effect of measurement (Goldberg 1969, 990, 991; Cuvaj 1968, 1104).

17. At one point Poincaré claims that if there were a deformation of bodies, like that postulated by Lorentz, it could be a deformation by any arbitrary continuous transformation and we still would not be able to recognize it (1913, 416). See Walter (2014a, 2014b) and Darrigol (1995, 2004) for up to date analyses of Poincaré's relation to the theories of Lorentz, Einstein, and Minkowski.

18. Sklar makes this point (1986, 14). Three different ways to defend relationalism have been discussed in the literature. Butterfield discusses the problems encountered in attempting to save the relationalist program by rewriting physical theories without referring to unoccupied spacetime points, using possible objects and possible configurations of objects instead. This move commits the relationalist to a possible world ontology (Butterfield 1984). Earman (1975) describes how it may be possible to eliminate the metric structure of spacetime by embedding it in a larger algebraic structure. Finally, Sklar opens the possibility of taking absolute acceleration and rotation as new primitive terms, and eliminating absolute spacetime (1974, 229–232); cf. the discussion by Friedman (1983, 232ff.). Nothing in Poincaré's philosophy of science rules out any of these options, but none of them have been shown to be satisfactory.

BIBLIOGRAPHY

Brunetière, Ferdinand. 1895. "Après une visite au Vatican." *Revue des Deux Mondes, 4e période* **127**: 97–118.
Butterfield, Jeremy. 1984. "Relationalism and Possible Worlds." *British Journal for the Philosophy of Science* **35**: 101–113.
Coffa, J. Alberto. 1986. "From Geometry to Tolerance: Sources of Conventionalism in 19th Century Geometry." In *From Quarks to Quasars: Philosophical Problems of Modern Physics*, edited by R. G. Colodny. Pittsburgh, PA: University of Pittsburgh Press, 3–70.
Cuvaj, Camillo. 1968. "Henri Poincaré's Mathematical Contributions to Relativity and the Poincaré Stresses." *American Journal of Physics* **36** (12): 1102–1113.

Darrigol, Olivier. 1995. "Henri Poincaré's Criticism of Fin de Siècle Electrodynamics." *Studies in History and Philosophy of Modern Physics* 26 (1): 1–44.
———. 2004. "The Mystery of the Einstein-Poincaré Connection." *Isis* 95 (4): 614–626.
DiSalle, Robert. 2006. *Understanding Space-Time: The Philosophical Development of Physics from Newton to Einstein*. Cambridge: Cambridge University Press.
Duhem, Pierre. (1905) 1954. *The Aim and Structure of Physical Theory*. Translated by P. P. Wiener. Princeton, NJ: Princeton University Press, 1954.
Earman, John. 1975. "Leibnizian Space-Times and Leibnizian Algebras." In *Historical and Philosophical Dimensions of Logic, Methodology and Philosophy of Science: Proceedings of the Fifth International Congress of Logic, Methodology and Philosophy of Science*, vol. 4, edited by R. E. Butts and J. Hintikka. Dordrecht: Reidel, 93–112.
Enriques, Federigo. (1911) 1991. "Principes de la Géométrie." In *Encyclopédie des Sciences Mathématiques Pures et Appliquées: Tome III (Premier Volume) Fondements de la Géométrie*, edited by J. Molk. Paris: Gauthier Villars et Teubner. Rpt. Paris: Jacques Gabay, 1–147.
Fine, Arthur. 1971. "Reflections on a Relational Theory of Space: Review of Adolf Grünbaum; *Geometry and Chronometry in Philosophical Perspective*." *Synthese* 22 (3/4): 448–481.
Folina, Janet. 1992. *Poincaré and the Philosophy of Mathematics*. London: MacMillan.
Friedman, Michael. 1983. *Foundations of Space-Time Theories: Relativistic Physics and Philosophy of Science*. Princeton, NJ: Princeton University Press.
———. 1999. *Reconsidering Logical Positivism*. Cambridge: Cambridge University Press.
Gardner, Martin. (1958) 1982. *Logic Machines and Diagrams*. Chicago: University of Chicago Press.
Giedymin, Jerzy. 1977. "On the Origin and Significance of Poincaré's Conventionalism." *Studies in History and Philosophy of Science* 8 (4): 271–301.
———. 1982. *Science and Convention: Essays on Henri Poincaré's Philosophy of Science and the Conventionalist Tradition*. Oxford: Pergamon Press.
———. 1991 "Geometrical and Physical Conventionalism of Henri Poincaré in Epistemological Formulation." *Studies in History and Philosophy of Science* 22 (1): 1–22.
Goldberg, Stanley. 1967. "Henri Poincaré and Einstein's Theory of Relativity." *American Journal of Physics* 35 (10): 934–944.
———. 1969. "The Lorentz Theory of Electrons and Einstein's Theory of Relativity." *American Journal of Physics* 37 (10): 982–994.
———. 1970. "Poincaré's Silence and Einstein's Relativity." *British Journal for the History of Science* 4: 73–84.
Gray, Jeremy J. 2008. *Plato's Ghost: The Modernist Transformation of Mathematics*. Princeton, NJ: Princeton University Press.
Grünbaum, Adolf. 1968. *Geometry and Chronmetry in Philosophical Perspective*. Minneapolis: University of Minnesota Press.
———. 1973. *Philosophical Problems of Space and Time*. 2nd ed. Dordrecht: D. Reidel.
Heinzmann, Gerhard, and David J. Stump. 2014. "Henri Poincaré." *The Stanford Encyclopedia of Philosophy* (Spring 2014 Edition), edited by Edward N. Zalta. http://plato.stanford.edu/archives/spr2014/entries/poincare/.
Holton, Gerald. 1964. "On the Thematic Analysis of Science: The Case of Poincaré and Relativity." In *Melanges Alexandre Koyre*. Paris: Herman, 257–268.
Ivanova, Milena. 2015. "Conventionalism, Structuralism and neo-Kantianism in Poincaré's Philosophy of Science" *Studies in History and Philosophy of Science Part B: Studies in History and Philosophy of Modern Physics*. http://dx.doi.org/10.1016/j.shpsb.2015.03.003.

Jevons, W. Stanley. 1869. *The Substition of Similars, the True Principle of Reasoning.* London: Macmillan.

Jevons, W. Stanley. (1958) 1974. *The Principles of Science: A Treatise on Logic and the Scientific Method.* London: Macmillan. Rpt. New York: Dover.

Johnson, Dale M. (1979) 1981. "The Problem of the Invariance of Dimension in the Growth of Modern Topology." *Archive for History of Exact Sciences* 20: 97–188.

Klein, F. 1921–1923. *Gesammelte Mathematische Abhandlungen.* Berlin: Verlag Julius Springer.

Lakatos, Imre. (1970) 1982. "Falsification and the Methodology of Scientific Research Programs." In *Criticism and the Growth of Knowledge*, edited by I. Lakatos and A. Musgrave. Cambridge: Cambridge University Press, 91–196.

Laudan, Larry. (1981) 1984. "A Confutation of Convergent Realism." *Philosophy of Science* 48. Rpt. in J. Leplin, ed., *Scientific Realism.* Berkeley: University of California Press, 218–249.

Lorentz, H. A. (1904) 1952. "Electromagnetic Phenomena in a System Moving with any Velocity Less than that of Light" *Proceedings of the Academy of Sciences of Amsterdam* 6: 809–830. Rpt. in Einstein et al., *The Principle of Relativity.* New York: Dover.

———. 1921. "Deux Mémoires de Henri Poincaré sur la Physique Mathématique" *Acta Mathematica* 38: 294–295.

Mach, Ernst. (1893) 1942. *The Science of Mechanics: a Critical and Historical Account of its Development.* Translated by T. McCormack. La Salle, IL: Open Court.

Massey, Gerald J. 1964. "Philosophy of Space." PhD diss., Princeton University, Princeton, NJ.

McCormmach, Russell. 1967. "Henri Poincaré and the Quantum Theory." *Isis* 58: 37–55.

Milnor, John. 1982. "Hyperbolic Geometry: The First 150 years." *American Mathematical Society Bulletin* 6: 9–24.

Nagel, Ernest. (1939) 1979. "The Formation of Modern Conceptions of Formal Logic in the Development of Geometry." In *Teleology Revisited and Other Essays in the Philosophy and History of Science.* New York, Columbia University Press, 195–259.

———. 1979. *The Structure of Science: Problems in the Logic of Scientific Explanation.* 2nd ed. Indianapolis, IN: Hackett Co.

Pap, Arthur. 1946. *The a priori in physical theory.* New York: King's Crown Press.

Picard, Émile. 1901. "Introduction Générale." In *Exposition Universelle Internationale de 1900 à Paris: Rapport du Jury International.* Paris: Imprimerie Nationale, 1–52.

Poincaré, Henri. 1897. "Réponse à quelque critiques." *Revue de Métaphysique et de Morale* 5: 59–70.

———. 1898. "On the Foundations of Geometry." *The Monist* 9: 1–43.

———. 1899a. "Des Fondements de la Géométrie: à propos d'un Livre de M. Russell." *Revue de Métaphysique et de Morale* 7: 251–279.

———. 1899b. "La logique et l'intuition dans la science mathématique et dans l'enseignement." *L'Enseignment Mathématique* 1: 157–162. Rpt. in *Oeuvres*, vol. XI, 129–133.

———. 1900. "Sur les Principles de Géométrie: Réponse à M. Russell." *Revue de Métaphysique et de Morale* 8: 73–86.

———. (1902a) 1913. "Sur la Valeur Objective de la Science." *Revue de Métaphysique et de Moral* 10, 263–293. Rpt. in Poincaré, *The Value of Science*, chap. X and XI, 321–355.

———. 1902b. "Les Fondements de la Géométrie (essay review of Hilbert, 1899)." *Bulletin des Sciences mathématiques*, 2nd series, 26: 249–272.

———. 1904. *Rapport sur les Travaux de M. Hilbert, professeur à l'Université de Goettingen, Présentés au troisième concours du prix Lobatchefsky.* Kasan: Imperial University Press.

———. (1905) 1970. *La Valeur de la Science.* Paris: Flammarion.

———. 1913. *The Foundations of Science: Science and Hypothesis, the Value of Science, Science and Method.* New York: The Science Press.

———. 1916–1956. *Oeuvres.* Paris: Gauthier-Villars.

———. 1963. *Mathematics and Science: Last Essays.* New York: Dover.

Paul, Harry. 1968. "The Debate over the Bankruptcy of Science in 1895." *French Historical Studies* 5: 299–327.

Richards, Joan L. 1994. "The Philosophy of Geometry to 1900." In *Companion Encyclopedia of the History and Philosophy of the Mathematical Sciences*, edited by I. Grattan-Guiness. London: Routledge, 913–919.

Rougier, Louis A. P. 1920. *La Philosophie géométrique de Henri Poincaré.* Paris: Alcan.

Russell, Bertrand. 1899. "Sur les Axioms de la Géométrie." *Revue de Métaphysique et de Morale* 7: 684–707.

Schlick, Moritz. 1935. "Are Natural Laws Conventions?" In *Readings in the Philosophy of Science*, edited by H. Feigl and M. Brodbeck. New York: Appleton Century Crofts, 181–188.

Sklar, Lawrence. 1974. *Space, Time, and Spacetime.* Berkeley: University of California Press.

———. 1986. *Philosophy and Spacetime Physics.* Berkeley: University of California Press.

Stump, David J. 1989 "Henri Poincaré's Philosophy of Science" *Studies in History and Philosophy of Science* 20: 335–363.

———. 1991. "Poincaré's Thesis of the Translatability of Euclidean and non-Euclidean Geometries." *Noûs* 25: 639–657.

———. 1996. "Poincaré's Curious Role in the Formalization of Mathematics." In *Henri Poincaré: Science and Philosophy, International Congress Nancy, France, 1994*, edited by J.L. Greffe, G. Heinzmann and K. Lorenz. Paris and Berlin: A. Blanchard and Akademie Verlag, 481–492.

———. 1998. "Poincaré, H.J." in *Routledge Encyclopedia of Philosophy*, vol. 7, edited by Edward Craig. London: Routledge, 478–483.

Torretti, Roberto. (1978) 1984. *Philosophy of Geometry from Riemann to Poincaré.* 2nd ed. Dordrecht: D. Reidel.

Tuller, Annita. 1967. *A Modern Introduction to Geometries.* Toronto: Van Nostrand.

Walter, Scott A. 2014a. "The Historical Origins of Spacetime." In *Springer Handbook of Spacetime*, edited by A. Ashtekar and V. Petkov. Berlin: Springer, 27–38.

———. 2014b. "Poincaré on Clocks in Motion." *Studies in History and Philosophy of Modern Physics* 47 (1): 131–141.

4 The Logical Empiricist or Positivist Engagement with A Priori Knowledge
Schlick, Reichenbach, Carnap, and Ayer

The standard account of a priori knowledge in the mid-twentieth century was that it simply does not exist, because all knowledge can be shown to be either empirical or simply a matter of definition. However, a look at the positions of individuals involved in the rejection of the a priori shows a rather nuanced and complicated story. Alberto Coffa, Michael Friedman, and others have argued that Schlick went through a neo-Kantian phase before becoming a strict empiricist, although this reading of Schlick has been challenged. Of course, in his later writings Schlick totally rejects the Kantian idea of the synthetic a priori, arguing that all synthetic statements are empirical and all a priori statements are tautologies. Before changing his view and adopting the standard account, Reichenbach considered a position on a priori knowledge that is a dynamic theory of the constitutive elements in science, as he forcefully distinguished between the constitutive role of the a priori and its claimed necessity. Carnap, in his later works, also shifted his position and ended up with a principle of tolerance that is more in keeping with the theories of the constitutive elements in science. Even Ayer, who as expected set out the most extreme form of empiricism, can be seen as acknowledging the special status of mathematics and logic and that their role is especially problematic for the empiricist.

Post Kant, you can define an empiricist as someone who rejects the existence of synthetic a priori truths. In outlining the Logical Empiricist or Positivist positions on a priori knowledge I will contend that the shifts in their positions shows ways in which the former a priori is still problematic and demonstrates the special status of mathematics and other fundamental principles as constitutive parts of physical theory. These were often treated as conventions, but of course the correct interpretation of conventionalism is itself contentious and also is broader in scope that the kind of theories that I am considering here. Yemima Ben-Menahem (2006) provides a recent account of conventions in the development of twentieth-century philosophy.

SCHLICK

Moritz Schlick wrote important books on Einstein's theory of relativity and on the theory of knowledge. As leader of the Vienna Circle, he made the case that a new scientific approach to philosophy can lead to genuine progress, claiming that recognizing the true nature of logic as formal (which is only possible once modern symbolic logic has been created) is the key to turning philosophers away from arid disputes and toward solvable problems, that will either be handed over to the empirical sciences such as psychology or shown to be solvable by logical analysis (Schlick [1930a] 1979). Schlick was decidedly more conservative and less politically engaged that Neurath or Carnap, and at least in his later writings he is more strictly empiricist.

Both Coffa and Friedman claim that Schlick went through an early neo-Kantian phase and that his final position is something less that full-blown empiricism, a point that is taken up in a recent article by Christian Bonnet and Ronan de Calan (2009, 120). In his early writings, Schlick considered the possibility that Einstein's principle of relativity might conflict with our a priori intuitions in a Kantian sense, but in his later works, Schlick rejects the Kantian a priori entirely. Friedman claims that Schlick recognized that claims about the geometry of space are far from being straightforwardly empirical (Friedman 1999, 60), but in saying this about the Logical Empiricists or Positivists generally (i.e., about Schlick, Reichenbach and Carnap), he only uses Reichenbach's early view as an example, so it is not clear how far he has made a case for Schlick. Furthermore, when arguing against conventionalism in his later writings, Schlick maintains that coordinating definitions, which are the basis of his interpretation of conventionalism, give laws of nature empirical content (Schlick 1935). I do not dwell on the controversy over whether Schlick is coming from a broadly Kantian background in his early writings and instead will focus on the form of empiricism that he develops in his later works.

To get a flavor of Schlick's later treatment of the a priori, consider his discussions of the a priori in his articles from the 1930s. In "Is There a Factual A Priori?" ([1930b] 1979) he is at pains to distance himself from the phenomenology associated with Husserl. Schlick rejects the idea that there is a synthetic a priori and objects to the claims of intuitive knowledge used by the phenomenologists. He seems to take a position that is in keeping with empiricism and to reject Kant, as well as Husserl. We can also consider Schlick's discussion of the status of logic in his article "Form and Content: An Introduction to Philosophical Thinking" ([1932] 1979). Schlick argues that the laws of logic (and by extension all the former a priori) are empty of content:

> If some day an astronomer should not find a planet at the place which he had calculated for its position, he would not think that the mistake lay in his using ordinary logic in his deductions, but he would know

that something was wrong with the hypotheses from which he deduced the position of the planet. (These hypotheses would concern the laws of motion, the initial position of the planets, the absence of disturbing influences etc.) A skeptic might object, that in principle the failure of the astronomer could be explained in two ways: (1) by inadequate hypotheses and (2) by inadequate logic. But the second explanation is impossible. It is based on the fundamental error that the calculation, as it were, adds something to the hypotheses, and that the result of the calculation is the product of two factors: initial assumptions and logical deduction. But this is not so. On the contrary, it is clear that the initial hypotheses alone determine the position of the planet. (Schlick 1932, 346–347)

As a minimum, Schlick holds that the truths of logic and mathematics are fixed and that we always look elsewhere for errors when we get a negative result, a view that is in keeping with Quine's idea of a hard core of scientific theories, that is, that there are theories or parts of theories that will never be given up. Going further than this, Schlick also gives principles of logic (and mathematics) a special status, namely, that they have no conditions for their validity. If logic is about anything, it is about the meaning expressed in our language. Schlick continues the quotation given earlier as follows:

The mathematical calculation by which the present position of the planet is 'deduced', from the general law does not do anything but show that the proposition about the particular place of the planet is already contained in the law; that is to say: that proposition is not a result of the law plus logic, but the law is an abbreviated way of asserting an indefinite number of propositions. One of these is picked out—that is all. Thus, if such a proposition is found to be false by observation, this proves that the law is false, it has nothing to do with logic. It must be clear by this time that the validity of logic (and mathematics) for the world does not presuppose anything about the world, not any 'rationality' of it, or whatever it may be called. It has nothing to do with any properties of the universe; it is concerned with the expression of facts by propositions (i.e., by other facts), and more particularly with the equivalence of different expressions. There are no conditions for the validity of logic. (Schlick 1932, 346–347)

It seems to me that Schlick leaves the validity of logic and mathematics a mystery here, but it is clear that he treats logic and mathematics in a manner in keeping with empiricism, that is, he takes them to be analytic. Schlick seems to think that the question of the validity of logic is a pseudo problem.

Another place to look at Schlick's views on the former a priori is to consider his interpretation of Poincaré. Here is Friedman's summary of Schlick's view, which we should remember does not agree with Friedman's own interpretation of Poincaré:

> Schlick's conception is derived both from Poincaré and from Hilbert. From Hilbert Schlick adopts the idea that the axioms of geometry "implicitly define" the primitive terms of that science. And this explains why the axioms of geometry are both nonempirical and conventional: alternative systems of geometry—Euclidean or non-Euclidean—simply count as different definitions of "point," "line," "between," and so on. Hilbert's view thus accounts for the nonempirical status of pure geometry. But we need to add Poincaré's ideas to account for applied geometry. Here Schlick reasons as follows: In applying such a purely formal system of implicit definitions to our actual experience of nature, no merely empirical considerations can force us to adopt one system rather than another; rather, only experience plus the requirement of overall *simplicity* of the laws of nature yield a determinate such system. (Friedman 1999, 64)

Schlick's interpretation of Poincaré leads directly to the empiricist version of conventionalism that I mentioned in Chapter 3. When they are interpreted as so-called implicit definitions conventions are not constitutive at all. Either they are mere nominal definitions and are hence analytic—this was dubbed Trivial Semantic Conventionalism in the literature—or else they are empirical. Indeed, all of the rest of theory is empirical, even if, as Friedman notes, Schlick's view is equivalent to the Duhem–Quine thesis or holism. Despite his alleged neo-Kantian roots, Schlick ends up as an empiricist who rejects Kant's synthetic a priori. We will also see that this stance leads him to reject totally the idea of a constitutive element in science. Indeed, he convinces Reichenbach to change his mind about the constitutive elements as well.

REICHENBACH

In his first book, *The Theory Relativity and A Priori Knowledge* (1920), Reichenbach takes a novel view of the Kantian a priori, which has recently been revived by Friedman. First, Reichenbach introduces a distinction between two meanings of *a priori* in Kant, which allows him to reject the necessity and universality of a priori claims so that he can account for the changes that have come about in our conception of geometry, stemming from the work of Hilbert and Einstein. On one hand, a priori knowledge is supposed to be eternally true and certain, and on the other hand, it is said to constitute the object of knowledge. Reichenbach says that we should give up the first meaning of the a priori from Kant but maintain the second, thereby allowing for a priori knowledge that has changed over time. While Kant took Euclidean metric geometry to be a priori in both senses, Reichenbach argues that in the light of Einstein's theory of relativity, metric geometry has changed but still remains constitutive in Kant's sense:

> Kant's concept of a priori has two different meanings. First, it means 'necessarily true' or 'true for all times' and secondly, 'constituting the concept of object.' The second meaning must be clarified. According to Kant, the object of knowledge, the thing of appearance, is not immediately given. Perceptions do not give the object, only the material of which it is constructed. Such constructions are achieved by an act of judgment. The judgment is the synthesis constructing the object from the manifold of the perception. ([1920] 1965, 48)

We can see here that Reichenbach's early view remains within the Kantian framework, in the sense that he retains Kant's idea that there are constitutive elements in science. Reichenbach focuses on principles of coordination, claiming that they have a unique epistemological status in scientific knowledge. As I mentioned in my introductory chapter, I now think that Reichenbach should be interpreted as having discovered the dynamic nature of the constitutive element in science. *Constitutive* is a broader term than *a priori* because there are theories of the constitutive elements of science, like Kant's, that take these elements to fixed and a priori, and there are theories of the constitutive elements that take them to have changed in different historical or other contexts. Pace Reichenbach, the focus should be on the constitutive elements, not on the a priori.

Friedman has shown that Reichenbach's views on the a priori changed after an exchange of letters with Schlick. Reichenbach gave up his defense of a modified Kantian a priori and embraced Schlick's empiricism. The essence of Reichenbach's early view can be seen in what follows, where he distinguishes himself from empiricism:

> [T]his view is distinct from an empiricist philosophy that believes it can characterize all scientific statements indifferently by the notion "derived from experience." Such an empiricist philosophy has not noticed the great difference existing between specific physical laws and the principles of coordination and is not aware of the fact that the latter have a completely different status from the former for the logical construction of knowledge. The doctrine of the a priori has been transformed into the theory that the logical construction of knowledge is determined by a special class of principle, and that this logical function singles out this class, the significance of which has nothing to do with the manner of its discovery and the duration of its validity. ([1920] 1965, 93–94, cited in Friedman 1999, 62)

In this early view, principles of coordination have a special status, and this is what Reichenbach means by constitution.[1] This special status is the key to maintaining a functional role for the a priori in Reichenbach's early works. Neither Schlick nor Einstein liked Reichenbach's proposal. Schlick sought to eliminate the a priori entirely through a strict empiricism that denies all synthetic a priori knowledge. After some correspondence, Schlick convinced Reichenbach to join him and change his view.

Einstein took a different position, a kind of epistemological holism influenced by Duhem (Howard 2010a, 2010b), but he was just as opposed to the notion of the a priori as was Schlick. Don Howard has several complaints with Friedman's theory of the dynamic a priori, but for now I will just mention one, namely, that Duhem has shown that the principles that are taken to be a priori and outside of the realm of empirical testing can in fact be changed, as long as we make adjustments in other parts of theory. Duhem complains that the conventionalists, specifically Le Roy and Poincaré, see certain principles as fixed, when in fact nothing is fixed (Howard 2010b, 338–339). Howard and Duhem are of course correct in saying that nothing can be permanently fixed, but the theory of constitutive elements in science does not commit one to saying that anything is objectively or permanently fixed. The most that can be said, without further argument, is that the principles are fixed in a specific context, that is, at a particular historical moment or even within a particular project with specific aims. All that you need to do is change your theory, however, to give some other element the role of the constitutive principle and make your original principle empirical. I argue that this contextualist reading of the constitutive elements in science does remain distinct from Duhemian and Quinean holism with its possible rearrangement of core and periphery, because the special epistemological status of the constitutive elements of science is maintained. Note that the status of the principles change in context, that is, empirical principles can become constitutive and vice versa. Of course, there may be a dispute among those holding constitutive theories on the question of whether or not anything is permanently in the constitutive role, so Le Roy and Poincaré may deserve some of the criticism. The most obviously area of contention will be mathematics, which may seem to be permanently constitutive and even a priori in the old sense, but I leave that discussion for Chapter 8. Here I simply say that from my pragmatic point of view, there is nothing that strikes me a permanently constitutive, even if there is always something or other that is playing a constitutive role in scientific theory.

Schlick convinced Reichenbach to give up his early view and to treat the principles of coordination as conventions (Friedman 1999, 62–68). These were further treated as linguistic, that is, a matter of definition, and one may freely choose principles of coordination limited only by consistency and the need for simplicity. Reichenbach's early view, which lies between Kantianism and strict empiricism, is important and is lost in his later work where he adopts Schlick's point of view. In particular, the special epistemological role that constitutive elements play in science is ignored. In his early writings, Reichenbach thought that the General Theory of Relativity had shown that metric conventionalism is false. Here is Friedman's explanation of a quote from Reichenbach:

> Reichenbach cannot accept Poincaré's (and Schlick's) conventionalism as a general philosophical doctrine about geometry as such, independent of any specific theoretical context, and for precisely, this reason Reichenbach explicitly rejects conventionalism in the "Introduction" to his book, "[M]athematicians asserted that a geometrical system was established according to conventions and represented an empty schema that did not contain any statements about the physical world. It was chosen on purely formal grounds and might equally well be replaced by a non-Euclidean schema. In the face of these criticisms the objection of the general theory of relativity embodies a completely new idea. This theory asserts simply and clearly that the theorems of Euclidean geometry do not apply to our physical space." (Reichenbach [1920] 1965, 3–4; Friedman 1999, 66–67)

Although the underlying mathematical structure of General Theory of Relativity does not have a determinate metric, metric is determined empirically in the General Theory of Relativity by the distribution of mass. This was not widely understood until the 1970s and 1980s, that is, until the relatively contemporary discussion of spacetime theories. I should mention, however, that I have found a similar idea from Federigo Enriques, always a critic of Poincaré's conventionalism, who writes in 1938 about "Poincaré's celebrated doctrine of conventions that we had the occasion to examine and refute in our book *Problems of Science* ([1906] 1914). The development of Einstein's theory of relativity has given the *coup de grâce* to this doctrine" (1938, 29), so at least some were aware of the conflict between Poincaré's conventionalism and the general theory of relativity.

In his post-1920 writings, Reichenbach sees geometry as relative, not conventional in Poincaré's sense as outlined in Chapter 3, rather in keeping with Schlick's interpretation of Poincaré in that a choice of a metric geometry matches whatever physical principles are needed to make the physical theory come out right when applied to the empirical reality. Geometry plus physics gives a total system that can be empirically equivalent to an alternative geometry if appropriate changes are made in the physics, an understanding that is expressed as Reichenbach's theorem Θ in the *Philosophy of Space and Time* ([1928] 1958, 33). Reichenbach claims shows that geometry is relative but not conventional, because he goes on to say that once definitions are fixed, we can get an empirical determination of the geometry of physical space, a result is due to Einstein's work on gravity. Reichenbach criticizes Poincaré for thinking that the question of which metric geometry applies to space remains open, even after fixing definitions. Thus, Reichenbach ends up advocating empiricism about the metric geometry, a point that is very clear in his comments on Einstein in the Schilpp volume (1970, 297).[2]

Despite his adoption of Schlick's empiricism, Reichenbach did not embrace Poincaré's geometric conventionalism in his later writings. Here is Reichenbach's critique of Poincaré's conventionalism from 1951, where he links a neo-Kantian view (here meaning that geometry remains a constitutive element in scientific theories) with Poincaré's conventionalism:

> On first sight this result looks like a confirmation of Kant's theory of space. If every geometry can be applied to the physical world, it seems as though geometry does not express a property of the physical world and is merely a subjective addition by the human observer, who in this way establishes an order among the objects of his perceptions. Neo-Kantians have used this argument in defense of their philosophy; and it was used in a philosophical conception called *conventionalism*, introduced by the French mathematician Henri Poincaré, according to whom geometry is a matter of convention and there is no meaning in a statement which purports to describe the geometry of the physical world.
>
> Closer investigation shows the argument to be untenable. Although every geometrical system can be used to describe the structure of the physical world, the geometrical system taken alone does not describe the structure completely. The description will be complete only if it includes a statement about the behavior of solid bodies and light rays. (1951, 133)

In his late writings, Reichenbach recognizes that geometric conventionalism based on holism, that is, on Schlick's interpretation of Poincaré, is insufficient. The behaviors of solid bodies and of light rays are empirical matters, so the geometry of spacetime can indeed be determined empirically, once definitions are fixed. In Reichenbach, we see a trajectory out of Kantianism and toward empiricism while we will see that Carnap, in one sense, takes the reverse course.

CARNAP

One of the things motivating Carnap early in his career was the rejection of metaphysics and of philosophical disputes in general. According to Carnap, the role of philosophy is analysis of the logical structure of language, and he adopts Russell and Frege's view that the grammar of ordinary language is misleading. Philosophy is able to correct mistakes and remove confusion by getting clear about language, while the rest is left to empirical science. Pseudo-statements, those that seem to say something but actually do not, are particularly dangerous. It is fine to create art or literature, but it is not right to claim that something can be learned about the world through metaphysics. Carnap's (1959) basic strategy is to show that what had formerly been considered metaphysical questions either cannot be raised at all, or else, they have a straightforward technical answer.

The arguments by Carnap against metaphysics may seem to be extremely abstract and even mean-spirited and intolerant. However, in Europe between the wars, nationalism was on the rise with devastating consequences. Nationalist movements used metaphysical concepts, such as the spirit of the German nation, so the principle of verification was originally a weapon of political struggle. In a similar vein, the claim of scientific objectivity and universality become the underpinnings of an internationalism that was the mainstay of the struggle against Fascism. It was not much later that Jewish professors were fired and a so-called German physics was established. The new kind of objectivity that the Logical Positivists and their cultural allies sought can be seen as a reaction to nationalism (Galison 1990).

It is interesting to discover the depth of Carnap's encounter with Heidegger (Friedman 2002, 2003). From historical documents, we now know that as a graduate student, Carnap attended a special seminar in Davos, Switzerland, that featured Heidegger and Cassirer. Heidegger and Carnap met at least twice for long discussions in cafes, and in their diaries, each said that they found the other very impressive. After the Davos seminar, Carnap studied Heidegger's work quite seriously. It is important that Carnap recognized that the fundamental disagreement between Heidegger and himself involves the place of logic. For Carnap, logic is a given, a place that we must start and taken to be universal. It is literally inconceivable to question logic, since all questioning (all thought) presupposes logic. Nevertheless, Heidegger thinks that the question of Being can never be properly answered if we start by taking logic for granted, rather, we must learn a new way of thinking. Their fundamental disagreement replicates and extends the dispute that developed within neo-Kantianism in German, with Heidegger and Carnap each taking the side of the school of neo-Kantianism in which they were trained. The "Marburg School" created a kind of logical idealism, interpreting the Kantian categories in a way that keeps philosophy compatible with the sciences. The "Southwest School" argued that formal logic cannot explain how the Kantian categories apply to our experience, so we have to start with the concrete given.

Late in his career Carnap adopts a view that is different from the strict empiricist rejection of the Kantian a priori. According to Carnap, natural science is to be represented in a formal language. In *The Logical Syntax of Language* ([1937] 1967), he distinguishes between L-rules and P-rules, that is, he divides a scientific language into a logical part and a physical part. The logical part is analytic and conventional, while the physical part is synthetic and empirical. Many different languages are possible, and the choice between them is practical, rather than cognitive. Carnap calls the choice of language an external question, and it is guided by the suitability of a language for a given purpose. There is no right or wrong language in an objective sense but, rather, only relative to our purposes. Carnap enunciated a principle of tolerance, famously saying that

> *[i]n logic, there are no morals.* Everyone is at liberty to build up his own logic, i.e. his own form of language, as he wishes. All that is required of him is that, if he wishes to discuss it, he must state his methods clearly, and give syntactical rules instead of philosophical arguments. ([1937] 1967, 52, emphasis in the original)

A change of language for Carnap amounts to a conceptual scientific revolution, with the change in language being a formal equivalent of a change in conceptual scheme or paradigm. Although he did not discuss historical changes in science, there is clearly a place for such revolutions in his outlook.

Even though Carnap does not link his analysis of science to the relativized a priori, several authors have recently noted the similarity between Carnap's views and those of Reichenbach and Friedman (e.g., Uebel 2012, § 3.2). As Paolo Parrini puts it, "Carnap can be seen as the philosopher who made the most significant contribution to the development of the Neo-empiricistic theory of the *relativized a priori*, by systematizing the standard *linguistic* version of this conception" (2009, 127, emphasis in the original). The main point here is that Carnap maintains a special role for the former a priori (and requires the analytic/synthetic distinction in order to do so). I completely agree with Parrini's assessment of Carnap's project:

> Today, we tend to see the strong attention given to the logical-formal aspect of the problem as a weakness of Carnap's approach. The trend towards logical formalism leads him to suppose that the analytic/synthetic distinction is purely a formal or logical distinction, and Quine showed that logically speaking all sentences of a given formal system are of equal value. Thus "Carnap's attempt further to characterize" some components of the system as semantical rules, meaning postulates or analytic sentences "ultimately amounts to nothing more than an otherwise arbitrary label." (Parrini 2009, 128)[3]

Carnap seems to have been endlessly optimistic about the power of formal methods, but his approach is overly formal and leaves itself open to the critiques developed by Quine and others.[4] Briefly summarizing the arguments here, in "Two Dogmas of Empiricism" (1953), Quine criticizes two ideas that are central to the program of the Logical Empiricists. First, he claims that the analytic/synthetic distinction cannot be drawn. Since this distinction is needed to even state what "modern empiricism" is (i.e., to reject the existence of synthetic a priori statements), Quine's critique is fundamental. Second, he rejects reductionism, the idea that every meaningful statement can be seen as a logical construction of observation predicates (which is equivalent to the principle of verification). In other words, he rejects the view that every meaningful sentence has some direct connection to experience. According to Quine, individual sentences are not directly verifiable or falsifiable, but rather, only a whole language is verifiable or falsifiable by experience (ironically Neurath's position is the same). Most of Quine's famous article concerns analyticity. The question Quine asks is how we know when a sentence is analytic, or alternatively, how we explain the fact that a sentence is analytic. There are various historical attempts to explain analyticity. The only plausible account says that analytic sentences are true by virtue of meaning; that is, 'All bachelors are male' is true because *bachelor* just means 'unmarried male'. Have we really explained why a sentence is analytic if we adopt this account? Quine claims that in order to understand how meaning can explain why a sentence is analytic, we must rely on a notion of synonymy. We can see this by contrasting sentences that are logically true (and hence analytic) with sentences that are analytic but not logically true. We can see immediately why the logical truth is analytic, it is a version of the sentence 'A = A', but not why the nonlogical truth is analytic. For example, I cannot immediately see why 'B = U' is analytic, unless I could substitute *unmarried male* for *bachelor*, then I would have the same idea as logical truth, 'U = U'. Quine's point is that we have not really *explained* what it means to be analytic, rather, we have simply relied on our previous understanding of synonymy in order to say that we have explained it. Since an understanding of notion of synonymy is necessary to explain analyticity, we really haven't got anywhere.

Quine next examines the question of whether the required notion of synonymy can be explained by an understanding of definition. Just looking at a dictionary will not furnish an explanation. Instead, we need an explication. But what does it mean to explicate a concept such as 'bachelor'? To provide the needed information, one has to already know which terms are synonymous, according to Quine. Another idea is that the notion of interchangeability will allow us to explain which terms are synonymous, but once again, Quine charges that we are explaining in circles. We explain analyticity in terms of meaning, meaning in terms of synonymy, synonymy in terms of words that are necessarily interchangeable, but all of these seem to simply be different ways of saying the same thing. Formal logic adds another problem to interchangeability given that many sentences have the same truth-value, but they are not all analytic.

Carnap approaches the problem of analyticity by adding semantic rules to a formal language (i.e., the meaning of terms is explicitly stated). However, these semantic rules again presuppose that we know when sentences are synonymous; otherwise, we could never formulate them. In summary, it seems impossible to define analyticity in formal terms; you can only define logical equivalence formally. Every explanation of why nonlogical statements are analytic presupposes that we know already what words are synonymous and that notion is dangerously close to simply being the same as analyticity. Finally, the verificationist theory of meaning could allow a distinction between analytic and synthetic statements—analytic statements are those that are always confirmed, no matter what experience we have. Carnap's *Aufbau* is the most developed attempt to make such a verificationist theory of meaning work, but it, too, relies on presupposed meanings, in particular, 'is at' is not defined. The alternative is holism: only a whole language is verifiable or falsifiable by experience, not individual sentences.

It is certainly possible to defend Carnap from this critique and many have done so (Grice and Strawson 1956; Creath 1991; Hunter 1995; Richardson 1997; Sober 2000; Parrini 2007). My own take on the matter is that we should move away from formal languages and trying to account for how meaning is specified and instead focus on scientific practice. I consider pragmatic theories of the constitutive elements in science in Chapter 5 and show that they come much closer to this ideal.

AYER

We can now turn to A. J. Ayer, who generally has not been treated in the recent literature on the Logical Empiricist or Positivist account of the a priori. Perhaps no one rejects the a priori more strongly than Ayer does in his *Language, Truth, and Logic* ([1935] 1950). Ayer is, of course, out of fashion, but since we have recently passed the centenary year of Ayer's birth, it may be appropriate to reconsider his views. He is credited with promulgating all sorts of misconceptions about the Vienna Circle and certainly did skew his reading of Logical Empiricism or Positivism toward British Empiricism while ignoring the neo-Kantian elements. To be fair, however, Ayer says up front in his "Preface to the First Edition" of *Language, Truth and Logic* that the views in the book derive from those of Russell and Wittgenstein, and through them from Hume and Berkeley, so he is consciously presenting an updated version of British empiricism (Ayer [1935] 1950, 31). Indeed, he even distances himself from positivism, referring specifically to Mach (Ayer [1935] 1950, 135), which is rather ironic given that *Language, Truth, and Logic* became a sort of textbook introduction to Logical Positivism. Despite these caveats, Ayer also says that he is closest to the view of the philosophers of the "Viennese Circle," especially Carnap, presumably meaning the early Carnap of the *Aufbau* (Ayer [1935] 1960, 32), so he cannot be completely blameless for giving the skewed reading that he does in fact give.

The problems with Ayer's connection to the Vienna Circle, however, only help to underscore the point that his treatment of the a priori is important, because it is in some ways the purest rejection of this element of knowledge. In particular, Ayer is not assuming a Kantian framework in setting up the issues. He argues for a strict empiricist removal of what had formerly been called the a priori, so that is that all knowledge of the world is empirical and that everything else reduces to definitions and is thus analytic. Ayer collapses the Kantian synthetic a priori, interpreting everything that had been in that category either as purely analytic or else as empirical and therefore not a priori at all.

It is noteworthy that Ayer devotes an entire chapter of *Language, Truth, and Logic* to the a priori. He has no problem advocating the view that all truths of science are contingent hypotheses, however, he does recognize a problem for his empiricist position when we consider the truths of mathematics and logic which must either be taken to be empirical as well in the manner of J.S. Mill, a position that Ayer rejects, or else be shown to be purely analytic and a matter of definition as in logicism ([1935] 1950, 73–74):

> The contention of Mill's which we reject is that the propositions of logic and mathematics have the same status as empirical hypotheses; that their validity is determined in the same way. We maintain that they are independent of experience in the sense that they do not owe their validity to empirical verification. We may come to discover them through an inductive process; but once we have apprehended them we see that they are necessarily true, that they hold good for every conceivable instance. And this serves to distinguish them from empirical generalizations. For we know that a proposition whose validity depends upon experience cannot be seen to be necessarily and universally true. (Ayer [1935] 1950, 75)

Note that the claims that Ayer makes in this quotation are perfectly in keeping with many of the theories of the constitutive element in science, given that he recognizes that the propositions of logic and mathematics have a special status. Of course, he thinks that they are tautologies and therefore analytic, but this does not mean that they are simply stipulative definitions, because, after all, they are useful in science and in everyday reasoning. Rather, Ayer thinks, we give them a special role by refusing to give them up, no matter what. Were we to find a putative refutation of a mathematical or logical principle, we would find some way to explain the refutation and show that the mathematical or logical principle is in fact true. Ayer treats logic and mathematics as even more certain than Quine's hard core of scientific theories, that is, he claims that we would never say that the mathematics or logic was wrong. "The principles of logic and mathematics are true universally simply because we never allow them to be anything else" (Ayer [1935] 1950, 77). Thus, it is ultimately a choice that we make, to always account for any apparent counter instance to a logical or mathematical principle by finding a way to maintain the principle. If one found that two times five did not equal ten, for example,

> one would say that I was wrong in supposing that there were five pairs of objects to start with, or that one of the objects had been taken away

while I was counting, or that two of them had coalesced, or that I had counted wrongly. One would adopt as an explanation whatever empirical hypothesis fitted in best with the accredited facts. (Ayer [1935] 1950, 75)

The question for us here is whether Ayer says more about the special status of logic and mathematics than Quine would, given that Quine's rejects that idea of constitutive elements in science. It turns out that Quine says ambiguous things about the status of logic,[5] but his final view is that mathematics lacks empirical content (Hahn and Schlipp 1998, 685). Indeed, that is his view in his last book (Quine 1995, 53). Ayer does have more to say about the status of logic however, because he does not merely claim that we would hold onto logic no matter what, but also claims that we would be contradicting ourselves if we tried to maintain that the rules of logic were wrong. Ayer continues the prior quote by saying, "And the reason for this is that we cannot abandon them without contradicting ourselves, without sinning against the rules which govern the use of language, and so making our utterances self-stultifying" ([1935] 1950, 77). Some caution is required here because in a note in the introduction to the second edition of *Language, Truth, and Logic*, Ayer renounces the view that a priori propositions "are themselves linguistic rules" ([1935] 1950, 17), so I take it that he finds another ground for why they are true and indeed necessary. The grounds for this necessity seems to be hypothetical, that is, if you accept the axioms a given formal system, then you must accept the theorems. But Ayer does not mention the fact that there must be some starting place, some reason why one is even applying the rule that make the hypothetical necessary.

CONCLUSION

The rejection of the Kantian synthetic a priori in the early twentieth century is more complex than it is often portrayed. Even while they rejected Kant, the Logical Empiricists may not have totally given up the special role that the former synthetic a priori played in knowledge, the role for what Alberto Coffa (1977) called "strange sentences." Ayer, coming out the of the tradition of British empiricism, had a different framework but still ends up with logic and mathematics playing a special role and not fitting easily into either analytic or empirical categories. Even if all knowledge is empirical, there are still significant differences between parts of our knowledge, given that some elements are constitutive of other elements. There is still some sense of the constitution of empirical knowledge even if the background principles that do the constitution are themselves ultimately empirical. These background principles do not have the same epistemic status as ordinary empirical theories.

Linguistic conventionalism, an influential misreading of Poincaré's space-time theories that interpreted spacetime conventionalism as a mere special case of the underdetermination of the meaning of primitive terms in all formal systems, seems to be the link between nineteenth-century developments in geometry and the received view of scientific theories developed by twentieth-century Logical Empiricists, but taking metamathematics as the source of conventionalism was a not correct. Furthermore, as recent scholarship has shown, it was applied geometry—mathematical physics—rather than the formal conception of pure mathematical theories that provided the exemplar for a new treatment of formerly a priori knowledge. Treating conventionalism as a metamathematical thesis led to the charge of "trivial semantic conventionalism"—a trivial linguistic thesis made Logical Empiricism vulnerable to the criticisms raised in Quine's "Two Dogmas of Empiricism" (1953). We will see in the next chapter that alternative conceptions of the a priori fare better than strict empiricism. Understanding the influential twentieth-century idea of an a priori that is changeable rather than necessary requires considering a range of options, with conventionalism and pragmatism falling between the extremes of idealism and strict empiricism.

NOTES

1. Friedman has recently given up basing his own dynamic view of the a priori on principles of coordination (2010, 777n253, 781n268). For more on the origins of Reichenbach's view, especially with regard to the principles of coordination, see Padovani (2011). Also see Oberdan (2009) on the exchange between Schlick and Reichenbach, especially on the issue of what gets constituted. Oberdan shows that Schlick especially objected to the idea of the constitution of objects by our scientific theories.
2. I thank an anonymous reviewer from Routledge for pressing me to clarify Reichenbach's view.
3. The quotations in my quotation are cited by Parrini as (Friedman 2001, 49; Parrini 2001, § 5).
4. In a recent article, Robert Hudson argues that Carnap cannot escape the critique of his principle of tolerance brought forth by Gödel, pace Ricketts and Friedman (Hudson, 2010).
5. See Haack (1996, 220) for a nice compilation of the things that Quine says about logic and the tension between them. Ben-Menahem (2006, 241) and Murphey (2011, 156) suggest possible solutions for Quine.

BIBLIOGRAPHY

Ayer, Alfred Jules. (1935) 1950. *Language, Truth, and Logic*. 2nd ed. New York: Dover.
Ben-Menahem, Yemima. 2006. *Conventionalism*. New York Cambridge University Press.

Bonnet, Christian, and Ronan de Calan. 2009. "Moritz Schlick: Between Synthetic *A Priori* Judgment and Conventionalism." In *Constituting Objectivity: Transcendental Perspectives on Modern Physics*, edited by M. Bitbol, P. Kerszberg, and J. Petitot. Berlin and New York: Springer, 117–126.

Carnap, Rudolf. (1937) 1967. *The Logical Syntax of Language.* Translated by A. Smeaton. London: Kegan Paul, Trench, Trubner, & Co.

———. 1959. "The Elimination of Metaphysics through Logical Analysis of Language." In *Logical Positivism*, edited by A. J. Ayer. New York: Free Press, 60–81.

Coffa, J. Alberto. 1977. "Carnap's *Sprachanschauung* circa 1932." In *PSA 1976*, edited by F. Suppe and P. Asquith. East Lansing, MI: Philosophy of Science Association, 205–241.

Creath, Richard. 1991. "Every Dogma Has its Day." *Erkenntnis* 35: 347–389.

Enriques, Federigo. (1906) 1914. *Problems of Science.* La Salle, IL: Open Court.

———. 1938. *La Théorie de la Connaissance Scientifique de Kant à nos Jours.* Paris: Hermann.

Friedman, Michael. 1999. *Reconsidering Logical Positivism.* Cambridge: Cambridge University Press.

———. 2001. *Dynamics of Reason: The 1999 Kant Lectures at Stanford University.* Stanford, CA: CSLI Publications.

———. 2002. "Carnap, Cassirer, and Heidegger: The Davos Disputation and Twentieth Century Philosophy." *European Journal of Philosophy* 10 (3): 263–274.

———. 2003. "A Turning Point in Philosophy: Carnap-Cassirer-Heidegger." In *Logical Empiricism: Historical and Contemporary Perspectives*, edited by P. Parrini, W. C. Salmon and M. H. Salmon. Pittsburgh, PA: University of Pittsburgh Press, 13–29.

———. 2010. "Synthetic History Reconsidered." In *Discourse on a New Method: Reinvigorating the Marriage of History and Philosophy of Science*, edited by M. Domski and M. Dickson. Chicago and La Salle, IL: Open Court, 571–813.

Galison, Peter. 1990. "Aufbau/Bauhaus: Logical Positivism and Architectural Modernism." *Critical Inquiry* 16: 709–752.

Grice, H. P., and P. F. Strawson. 1956. "In Defense of a Dogma." *The Philosophical Review* 65 (2): 141–158.

Haack, Susan. 1996. *Deviant Logic, Fuzzy Logic: Beyond the Formalism.* Chicago: University of Chicago Press.

Hahn, Lewis Edwin, and Paul Arthur Schilpp, eds. 1998. *The Philosophy of W. V. Quine.* 2nd expanded ed. The Library of Living Philosophers, vol. 18. La Salle, IL: Open Court.

Howard, Don A. 2010a. "Einstein's Philosophy of Science." *The Stanford Encyclopedia of Philosophy* (Summer 2010 Edition), edited by Edward N. Zalta. http://plato.stanford.edu/archives/sum2010/entries/einstein-philscience/.

———. 2010b. "'Let Me Briefly Indicate Why I Do Not Find This Standpoint Natural.' Einstein, General Relativity, and the Contingent A Priori." In *Discourse on a New Method: Reinvigorating the Marriage of History and Philosophy of Science*, edited by M. Domski and M. Dickson. Chicago and La Salle, IL: Open Court, 333–355.

Hudson, Robert. 2010. "Carnap, the Principle of Tolerance, and Empiricism." *Philosophy of Science* 77 (July): 341–358.

Hunter, Geoffrey. 1995. "Quine's Two Dogmas of Empiricism.'" *Philosophical Investigations* 18 (4): 305–328.

Murphey, Murray G. 2011. *The Development of Quine's Philosophy.* New York: Springer.

Oberdan, Thomas. 2009. "Geometry, Convention, and the Relativized Apriori: The Schlick—Reichenbach Correspondence." In *Stationen. Dem Philosophen und*

Physiker Moritz Schlick zum 125. Geburtstag, edited by F. Stadler, H.J. Wendel, and E. Glassner. Vienna: Springer, 186–211.

Padovani, Flavia. 2011. "Relativizing the Relativized A Priori: Reichenbach's Axioms of Coordination Divided." *Synthese* 181: 41–62.

Parrini, Paolo. 2001. "The 'Dogma' of Analyticity Fifty Years After." In *Logic and Metaphysics*, edited by M. Marsonet and M. Benzi. Genova: Name edizione, 103–131.

———. 2007. "The 'Two Dogmas of Empiricism' 50 Years On." *Diogenes* 54 (4): 91–101.

———. 2009. "Carnap's Relativised A Priori and Ontology." In *Constituting Objectivity: Transcendental Perspectives on Modern Physics*, edited by M. Bitbol, P. Kerszberg and J. Petitot. Berlin and New York: Springer, 127–143.

Quine, Willard Van Orman. 1953. "Two Dogmas of Empiricism." In *From a Logical Point of View*. Cambridge, MA: Harvard University Press, 20–46.

———. 1995. *From Stimulus to Science*. Cambridge, MA: Harvard University Press.

Reichenbach, Hans. (1920) 1965. *The Theory of Relativity and A Priori Knowledge*. Berkeley: University of California Press.

———. (1928) 1958. *Philosophy of Space and Time*. New York: Dover.

———. 1951. *The Rise of Scientific Philosophy*. Berkeley: University of California Press.

Richardson, Alan. 1997. "Two Dogmas about Logical Empiricism: Carnap and Quine on Logic, Epistemology, and Empiricism." *Philosophical Topics* 25 (2): 145–168.

Schilpp, Paul Arthur, ed. 1970. *Albert Einstein, Philosopher-Scientist*. 3rd ed. The Library of Living Philosophers, vol. 7. La Salle, IL: Open Court.

Schlick, Moritz. (1930a) 1979. "The Turning Point in Philosophy." In *Moritz Schlick Philosophical Papers vol. II (1925–1936)*, edited by H. L. Mulder and B. F. B. Van de Velde-Schlick. Dordrecht: D. Reidel, 154–160.

———. (1930b) 1979. "Is There a Factual A Priori?" In *Moritz Schlick Philosophical Papers Vol. II (1925–1936)*, edited by H. L. Mulder and B. F. B. Van de Velde-Schlick. Dordrecht: D. Reidel, 161–170.

———. (1932) 1979. "Form and Content: An Introduction to Philosophical Thinking." In *Moritz Schlick Philosophical Papers Vol. II (1925–1936)*, edited by H.L. Mulder and B.F.B. Van de Velde-Schlick. Dordrecht: D. Reidel, 285–368.

———. 1935. "Are Natural Laws Conventions?" In *Readings in the Philosophy of Science*, edited by H. Feigl and M. Brodbeck. New York: Appleton Century Crofts, 181–188.

Sober, Elliott. 2000. "Quine's Two Dogmas." *Proceedings of the Aristotelian Society* 74 (Suppl.): 237–80.

Uebel, Thomas. 2012. "Vienna Circle." *The Stanford Encyclopedia of Philosophy* (Summer 2012 Edition), edited by Edward N. Zalta. http://plato.stanford.edu/archives/sum2012/entries/vienna-circle/.

5 Alternative Conceptions of the A Priori
Cassirer, Lewis, and Pap

The Logical Positivist account of the a priori was not the only one in mid-twentieth century philosophy of science given that the neo-Kantians, pragmatists, and those influenced by them had alternatives. While rejecting the traditional view of a priori knowledge as certain and fixed, these accounts acknowledged a special role for what had been considered a priori knowledge. In that sense, they are versions of the theories of the constitutive elements in science that we have been following. I present the ideas of Ernst Cassirer, C.I. Lewis, and Arthur Pap in order to compare them to the contemporary theories of the constitutive elements in science and to show how they and John Dewey presented an alternative to Logical Positivism in twentieth-century philosophy of science.

ERNST CASSIRER

As a neo-Kantian, Cassirer can be seen as a transitional figure in the development of alternative theories of constitutive elements in science, or in what was taken to be the a priori. As the neo-Kantians worked to assimilate new developments in mathematics and science into a Kantian philosophical framework, they had to revise Kant's theory of the a priori, especially as it pertains to geometry. Kant said that there is an independent a priori faculty of sensibility, given our a priori intuitions of space and time, which grounds the certainty that he finds in our knowledge of Euclidean geometry and in arithmetic. The neo-Kantians eliminated the faculty of sensibility, leaving an a priori grounded in the understanding. For our purposes, we can say that the neo-Kantians maintained the existence of a priori elements in science but that they gave up on one of the sources of the a priori found in Kant.

Considerable difference of opinion about Cassirer's views on the a priori appear in the literature. Friedman interprets Cassirer as holding that there are fixed and universal elements in science rather than a dynamic a priori (Friedman 2000, 115ff.), while Richardson (1998, chap. 5) and Ryckman (2005, chap. 2) see him as holding a version of the dynamic a priori (Heis 2014a, 11). As we will see, Pap also understood Cassirer as advocating the idea that the a priori can change, and Ferrari has argued persuasively that Friedman underestimates Cassirer's contribution to the idea of the dynamic or historical a priori (Ferrari 2009, 300; 2012, 22). Fortunately, a recent paper by Jeremy Heis (2014a) has shed light on these issues, showing that Cassirer holds that there are both universal and fixed a priori elements in science, as well as historically changing a priori elements. Therefore, it is quite possible to read him as both the grandfather of the dynamic a priori and at the same time a defender of fixed and universal a priori elements in science.

Cassirer is a complex figure whose philosophy encompassed an extremely broad array of topics (Friedman 2011), but two elements of his thought will help us understand the position that he takes on the a priori. First, Cassirer sees progress in knowledge as resulting from a shift away from thinking in terms of what he calls substance-concept (*Substanzbegriff*) and toward what he calls function-concept (*Funktionsbegriff*), which is the main idea in his major work in the philosophy of science, *Substance and Function* (1923):

> Cassirer means this contrast to cover a number of different interrelated epistemic, logical, and metaphysical contrasts. Of these many uses, the most fundamental use that Cassirer makes of "Substanzbegriff" and "Funktionsbegriff" is epistemological and Kantian: it contrasts philosophical views that overlook the epistemic preconditions of various kinds of knowledge, with those that recognize the "functions" [or "preconditions"] that make certain kinds of knowledge possible. To oppose the point of view of "Substanzbegriff," then, is to oppose various forms of epistemological atomism: the view that certain kinds of knowledge (be they scientific concepts, experiences, or measurements) could be acquired all by themselves, without any other epistemic conditions (Heis 2014a, 11; also see Heis 2014b).

Following the lead of his fellow neo-Kantian, Hermann Cohen, Cassirer takes the existence of scientific knowledge as a given.[1] The task of philosophy is to explain how science is possible and to find the fundamental conditions of its possibility. Given the changes that took place in nineteenth- and twentieth-century mathematics and science, Cassirer responds by looking for deeper, more general, and thus still universal principles that underlay both the old and the new sciences, while acknowledging that some things that are a priori in Kant's philosophy have changed.

It is the fundamental contrast between substance-concept and function-concept that Cassirer sees as the motivation for looking for deeper, more general principles that form the basis of the practice of science.[2] To take a concrete example, Lovrenov (2006) argues that for Cassirer, metric geometry is not a priori, but the group theory that underlies it is. Thus, in Cassirer, as in Poincaré, there is a more general, abstract element in science that remains invariant and forms the basis of the alternative metric geometries.

A change from Euclidean to non-Euclidean geometry in physics would then be seen as a change from one group of displacements to another group. Group theory would remain constant while metric geometry would change. Ferrari makes the same point as follows:

> Such forms of "functional thinking" are only graspable as a priori forms—hence the a priori of space contains "no assertion regarding a single determinate structure of space in itself," rather it concerns "only the function of 'spatiality' in general, wholly distinct of its more specific determination" (Cassirer, 1987, p. 93). The question of the "real", Euclidean or non-Euclidean structure of space therefore loses "all relevance", as the new task at hand is to ascertain the a priori validity of spatiality as a general relation and not to define its empirical measurement. (2012, 22)

In other words, we are unable to specify, a priori, whether space is Euclidean or non-Euclidean, but we can specify the concept of spatiality as a general relation that underlies both metric geometries. While the underlying functional element may remain unchanged, the other a priori elements change as knowledge progresses. Ferrari shows how Cassirer confirms this viewpoint in a letter to Schlick, who had written a critical review of Cassirer's book on Einstein (Schlick 1921):

> As I can see, the point of disagreement between us lies in the notion of a priori, which I understand in a different way to you: namely, not as a steady and definitively established complex of material intuitions or concepts, but only as a function, which is determined according to a law and therefore it remains the same regarding its direction and its form; nevertheless it can assume the most various developments in the progress of knowledge. I would like to consider as a priori valid in a rigorous sense only the idea of 'unity of nature', that is the lawfulness of experience in general, or put in a brief formulation: the 'uniqueness (Eindeutigkeit) of coordination. (Cassirer, 2009, 50–51, cited in Ferrari 2012, 22).

The form or structure remains the same, but the concrete content can change in the development of science. As Ryckman points out, the 1921 review can be seen as a watershed in the philosophy of science given that Schlick's view of the a priori won out over that of Cassirer (Ryckman 2003, 161). In particular, Schlick rejected Cassirer's contention that the fundamental principles of science are synthetic, holding rather that they are merely verbal stipulations and thus analytic. As we saw in Chapter 4, this is the crux of the empiricist position: that there are no synthetic a priori truths.

Besides the move to functional or more abstract thinking, which allows a two-tier version of the a priori in Cassirer, with the fundamental level remaining constant and the other level changing over time, the second main idea motivating Cassirer's view of the a priori is that science is objective. As Heis emphasizes, Cassirer thinks that there must be some fixed elements in science in order to show continuity across scientific revolutions:

> The objectivity of the revision or replacement of the highest laws in a physical theory is a particularly pressing issue, since (following Duhem) Cassirer allows that there could be two distinct theories equally compatible with the empirical evidence. For this reason, Cassirer argues that there are supreme "laws" or norms that guide the formation and selection of new physical theories. Science can remain objective, then, even as theories come and go, since there are permanent laws shaping this process of theory formation and selection. (2014a, 12)

Cassirer looks to the functional elements in the a priori as forming the basis of all scientific theories and as being constant even when there are genuine scientific revolutions. These elements are still a priori in the old sense, though there is a difference from Kant, in that they are regulative instead of constitutive, in the Kantian senses of these terms.[3] However, my point here can be made without getting into the technical details of Kant's distinction. In Cassirer's a priori, some elements are dynamic or changing, while other elements are fixed and universal, but both kinds of elements are constitutive in the sense of being necessary preconditions to the possibility of science.

Cassirer believes that the universal elements are the basis for the objectivity of science. My own view is that it is doubtful whether there are in fact any such universal elements in science and furthermore, it is unnecessary to have fixed or universal elements in order to defend the objectivity of science. The full argument for my view will have to wait for later chapters, but here I would like to raise one objection to the idea of universal and fixed elements. As Heis explains Cassirer's view, he notes that Cassirer should not be taken to be saying that we have certain knowledge of the fundamental unchanging elements of science, which would indeed be a difficult position to defend. Rather, we are supposed to be converging on the fundamental ideas, but we may be wrong about them at the current stage of development of science:

> As Cassirer put it in SF [1910, 1923] p. 269 "at no given stage of knowledge can this goal [of identifying the ultimate invariants] be perfectly achieved," but instead these absolute a priori elements remain "as a demand" that guide the "continuous unfolding and evolution" of our physical theories . . . though concepts like <space> and <time> are a priori in this absolute sense, it is still the case that we can be radically mistaken about their nature. Although these concepts remained invariant during the switch from Newtonian to relativistic physics, space went from being thought of as a 3-dimensional Euclidean container independent of time and the distribution of matter,

to a 4-dimensional Riemannian spacetime whose structure evolves dynamically. (Heis 2014a, 15)

Cassirer's position here is untenable. What we actually use in science is our current understanding of concepts, not what the concepts really are in some Platonic heaven or what they will be at the hypothetical end of inquiry. The understanding of the concepts that we actually use in science do change from one historical period to another and therefore the supposedly universal and fixed elements of the a priori vanish. If we look at science as it is actually practiced, as I think that we must, there are no universal and fixed concepts in science, given that the understanding of the fundamental concepts changes over time.

I cannot, of course, prove that there never will be any universal and fixed concepts in science, given that I would have to refute an unlimited number of cases. Nevertheless, the history of science seems to show that concepts change in ways that seem unimaginable to those who are working with them prior to a revolution. Indeed, if we take fallibilism seriously, any of our current beliefs could turn out to be wrong, which implies that any of them could change in the long-term course of inquiry. Fallibilism gives us good reason to doubt that there ever were or ever will be any universal and fixed elements in science. We will see that pragmatism gives us a way to live without them.

C. I. LEWIS

C. I. Lewis's classic article "A Pragmatic Conception of the A Priori" (1923) lists three areas of a priori knowledge. The first is logic, which he calls "the traditional example of the a priori par excellence" (Lewis [1923] 1970, 231). The second, which he takes to be rather trivial, are propositions that are true by definition. Finally, Lewis lists the place of the a priori in science, which he says is "greater than might be supposed" (Lewis [1923] 1970, 234). The role of the pragmatic a priori in science is actually what is new and distinctive in his theory and the part that is a theory of constitutive elements in science.[4] Therefore, my focus is on this last category listed by Lewis, but we might first consider the other two.

Lewis says that the laws of logic "are independent of experience because they impose no limitations whatever upon it. They are legislative because they are addressed to ourselves—because definition, classification, and inference represent no operations of the objective world, but only our own categorical attitudes of mind" ([1923] 1970, 232). We might now say that logic is about language rather than about mind, but the point is the same. The laws of logic do not control things in the world but, rather, control the way that we think about things in the world or the way that we express ideas about things in the world. Logic is included in all thought so it cannot be separated from our thought, which means that anytime you are thinking, you are following the laws of logic. The idea behind logic being a priori is that it is so fundamental that it cannot be based on anything else. Logic is self-justifying because the principles of logic must be used in its own justification. However, the laws of logic are ultimately pragmatic, according to Lewis. There are alternative logics (Lewis's own groundbreaking systems of modal logic are an example), which shows that there is no single universal logic on which everyone must agree. Furthermore, Lewis argues that there are no facts of the matter about logic by which one can decide on the alternatives. Rather, one must look to pragmatic considerations of utility, convenience, and the like to decide on questions in logic. Thus, logic functions as a priori knowledge but is not immutable and fixed, according to Lewis. Both logic and mathematics are a priori and conventional in the sense that they are created from a series of definitions and chosen axioms and both can be applied and become part of an empirical science.

Regarding a priori propositions that are "true by definition" (Lewis puts the phrase in quotation marks), he seems to think that they are unproblematic and even trivial. Sentences are true by definition when concepts mentioned in them are related. For example, 'all bachelors are unmarried' is true by definition, because the concept of bachelor includes the idea of being unmarried. There is no right or wrong way to define terms in an absolute sense, but definitions must pass the pragmatic test of usefulness. "If experience were other than it is, the definition and its corresponding classification might be inconvenient, fantastic, or useless, but it could not be false" (Lewis [1923] 1970, 233).

Quine's critique of the analytic/synthetic distinction shows that the idea of sentences that are true by definition is not unproblematic. After all, the whole line of argument against the analytic/synthetic distinction in "Two dogmas of Empiricism" starts with a critique of the idea of sentences that are true by definition (Quine 1953). I neither criticize nor defend Lewis here, but rather just point out that the notion of sentences that are true by definition is more problematic than Lewis thinks. Throughout his career Lewis maintained the analytic/synthetic distinction, thinking that it was fundamental, which is why Lewis is explicitly mentioned, along with Carnap, as a target in "Two Dogmas of Empiricism." Quine thus presents a strong challenge to this part of Lewis's pragmatic theory of the a priori; however, we still have what Lewis calls the a priori in logic and the a priori in science. We can say that these are constitutive even if Quine is right about the analytic/synthetic distinction, which is enough to make Lewis's theory of considerable importance to us here.

THE A PRIORI IN PHYSICS

The laws of physics present the most interesting part of Lewis's pragmatic theory of the a priori. Along with definitive concepts which form the basis of any natural science and are hence a priori, Lewis accepts the Kantian point that certain principles constitute a science in the sense that they are necessary preconditions. "The fundamental laws of any science—or those treated as fundamental—are a priori because they formulate just such definitive concepts or categorical tests by which alone investigation becomes possible" (Lewis [1923] 1970, 234–235). The fundamental laws are examples of necessary preconditions, so we see that Lewis has a theory of the constitutive elements in science though as I have argued earlier, it seems unnecessary and confusing to call these elements a priori. Nevertheless, Lewis expresses the important idea that the constitutive elements make a science possible. Although the constitutive elements function as a priori statements, they are not fixed and must only meet a pragmatic test of usefulness. It is possible that different fundamental laws could play the same role. We will see that both Hacking and Friedman adopt similar language.

Lewis uses a long quote from Einstein on the definition of simultaneity as his only example of a constitutive element in science. The fact that he has only one example and that it turns out to be controversial considerably weakens Lewis's argument for the pragmatic a priori in science. In the long quotation, we learn that the principle that light always travels at the same speed in a vacuum is a stipulation that cannot be justified either empirically or by direct inspection. Such a stipulation is necessary in order to define simultaneity, given that trying to define the principle empirically leads to a circle. To set up his explanation of this point, Einstein asks how it is possible to judge whether lightning strikes at two separate locations are simultaneous. The obvious suggestion is to put an observer exactly midway between the two locations and have them judge whether or not the flashes of light are simultaneous by observing whether they arrive at the same time. However, for this suggestion to work, it has to be the case that light travels at the same speed from each location to the observer: "But an examination of this supposition would only be possible if we already had at our disposal the means of measuring time. It would thus appear as though we were moving here in a logical circle" (Einstein 1920, 27; quoted in Lewis [1923] 1970, 235).[5] We cannot measure time without a definition of simultaneity, but that is the very thing that we are trying to define. To check whether light travels at the same speed from each distant location to the observer, we need a way to measure the speed of light, that is, the distance traveled over a given time, and to make this measurement, we need know the time that each strike of lightning occurred and when it reached the observer. However, we were trying to define simultaneity, and by this empirical method, we have to already know that the time of the lightning strikes is the same.

Einstein's solution to this problem, as is well known, is to stipulate that light always travels at the same speed in a vacuum, independent of a frame of reference, and this stipulation is the prime example of what Lewis calls an a priori principle in natural science, what I would call a constitutive principle. As we have seen, Einstein argues that it is impossible to define simultaneity empirically without arguing in a circle. Furthermore, we have no direct insight into the principle that would justify it a priori in the traditional sense. Just as in Poincaré's argument about the conventionality of metric geometry, both an empirical determination and a traditional a priori determination of the principle have been ruled out, so the principle is an interesting example of a constitutive element in science because it is a principle with a strange epistemological status, being neither empirical nor a priori. After the page-long quote from Einstein in "The Pragmatic Theory of the A Priori," Lewis remarks,

> As this example from the theory of relativity well illustrates, we cannot even ask the questions which discovered law would answer until we have first by *a priori* stipulation formulated definitive criteria. Such concepts are not verbal definitions, nor classifications merely; they are themselves laws which prescribe a certain uniformity of behavior to whatever is thus named. Such definitive laws are *a priori*; only so can we enter upon the investigation by 'which further laws are sought. Yet it should also be pointed out that such *a priori* laws are subject to abandonment if the structure which is built upon them does not succeed in simplifying our interpretation of phenomena. (Lewis [1923] 1970, 236)

Lewis here again emphasizes the idea of the a priori as a precondition and this Kantian sounding standard for the a priori is central to Lewis's theory. Indeed, I believe that his theory of the a priori in science could be correctly described as the view that the fundamental presuppositions of an empirical theory are to be taken to be a priori, as long as there is no straightforward way to verify the presuppositions empirically. It is taken for granted that there will be no way to verify a presupposition by intuition or direct inspection, as we might expect in a classical rationalist view of a priori knowledge.

Before discussing the conventionality of simultaneity, let us get clear that the relativity of simultaneity and conventionality of simultaneity are not the same thing. Indeed, they are introduced in two separate chapters of Einstein's popular book on relativity that Lewis quotes (Einstein 1920). Reichenbach and some others have not clearly distinguished them and this can be confusing, not the least because the conventionality of simultaneity is controversial, while the relativity of simultaneity is not. The point of the relativity of simultaneity is that whether or not two events are simultaneous depends on your reference frame. To take Einstein's simple example, suppose that lightning strikes at two locations, A and B, along a railroad embankment. To an observer on the embankment at the midpoint between A and B, suppose that the flashes of light arrive at the same time. In that case, we say that the lightning strikes are simultaneous for the observer at the embankment. Suppose further that a train is passing at that moment and that an observer on the train passes the observer on the embankment and sees the lightning strikes. The train is racing toward A and away from B, so in the time that the flash of lightning at A reaches the midpoint, the train will have moved forward. Therefore, although the observer on the train will see the flash of light at the same time, she will interpret it differently from the observer on the platform, since the light has to travel a shorter path to her. So, for the observer on the train, the flashes of light are not simultaneous—the light from A appears before the light from B (Norton 2014). The conclusion to draw is that whether or not two events are simultaneous depends on your reference frame—this is the relativity of simultaneity.

The conventionality of simultaneity, on the other hand, refers to the fact that one must make a more or less arbitrary assumption in defining simultaneity, given that one must stipulate that the one-way speed of light is always the same. In the case of the simultaneous events mentioned above, we must stipulate that light travels at the same speed from A to the midpoint as it does from B to the midpoint. Reichenbach seems to run together the ideas of the relativity and conventionality of simultaneity when he says, "The word 'relativity' is intended to express the fact that the results of measurements depend on the choice of coordinative definitions" (1958, 15). In this passage Reichenbach clearly associates the word *relativity* with the choice of coordinative definitions, while in fact, this is what is meant by the conventionality, not relativity of simultaneity. However, his discussion of Einstein's example and the argument that the definition of simultaneity is a coordinative definition does not use the term *relativity* (Reichenbach [1928] 1958, 126–127).

It appears that what Lewis needs to make his point about the pragmatic a priori in science is the conventionality of simultaneity, and since that is a controversial claim, it seems that his only example of the pragmatic a priori in science may be undercut. There are two basic ways of arguing against the conventionality of simultaneity in the philosophy of science literature: slow clock transport and Malament's theorem. Although Janis makes a very good case that the issue is still an open one in his recent survey of the literature (Janis 2014), it certainly weakens Lewis's argument to base everything on one example that may not even work. Without trying to settle the issue of whether the definition of simultaneity is conventional here, I can only say that it is too bad that so much is riding on this one case for Lewis. It is true that the definition of simultaneity is a necessary precondition to further inquiry, so the definition plays a special role, however it may not be the kind of presupposition that is needed to make the case for constitutive elements in science.

It could be said that Lewis overemphasizes the conceptual, given that Einstein's stipulation is more than just a definition of terms, rather it tells us about a physical process and defines a set of practices, namely how to measure the speed of light. It is striking that while his example goes beyond the purely conceptual, Lewis states that the a priori is conceptual in nature—that it only has to do with defining terms. I suggest that the example that he borrows from Einstein belies Lewis's explicit statement of the a priori as conceptual. It is how you treat these kinds of principles that have the special status of looking like empirical laws but actually being necessary preconditions that is the key point.

LEWIS'S THEORY OF KNOWLEDGE

Lewis's pragmatic theory of the a priori is imbedded in a general theory of knowledge, or as he sometimes says, a theory of experience, which is set out in *Mind and World Order* (Lewis 1929). According to Lewis, there are two elements of experience, the concept and the given in sense perception: "Knowledge of objects, then, knowledge of the real, involves always two elements, the element of the given and ineffable presentation, and the element of conceptual interpretation which represents the mind's response" (Lewis 1929, 143). Thus, there are actually two stages in which the mind is active, in forming the concept and in conceptual interpretation of the given according to concepts. It is only in the abstract that we can separate these elements of experience according to Lewis, because we never experience the given as a sense data or qualia; rather, we experience an object. We do not see a red patch; rather, we see an apple. Thus, Lewis is quite far from the Logical Positivist or Empiricist view of experience and closer to the view of critics of these movements. Indeed, some of Lewis's statements are reminiscent of Quine's web of belief, with its periphery and core:

> I would emphasize the fact that the whole body of our conceptual interpretations form a sort of hierarchy or pyramid with the most comprehensive, such as those of logic, at the top, and the least general, such as 'swans,' etc., at the bottom. . . . The higher up a concept stands in our pyramid, the more reluctant we are to disturb it, because the more radical and far-reaching that results will be if we abandon the application of it in some particular fashion. (Lewis 1929, 305–306)

Nevertheless, Lewis insists that the given can be separated from experience in analysis. Experience is defined as bringing these theories and concepts to the world and interpreting whatever is given.

Lewis recapitulates the key elements of his theory of the a priori in chapter 8 of *Mind and World Order*, "The Nature of the A Priori, and the Pragmatic Element in Knowledge." He uses the same long quote from Einstein that he used in his article "The Pragmatic Theory of the A Priori" to introduce the a priori in physical theory (Lewis 1929, 254–256), and again lists logic and definitions as the other two examples of a priori truths. Mathematics comes in for a more extended treatment here, and he is somewhat more circumspect about logicism, listing in a footnote some reservations about what Russell and Whitehead have shown in *Principia Mathematica* (Lewis 1929, 244). For example, he raises a question of the status about the axiom of infinity, recognizing that it may not seem to be a principle of logic.

In the end, Lewis holds that pure mathematics is in between logic and the empirical application of mathematics (Lewis 1929, 245–246). Pure mathematics is only limited by the requirement of consistency, while applications must meet the pragmatic test of being useful.

The rest of the a priori, and especially the a priori aspects of physical theory, are further constrained by pragmatic considerations. Lewis insists that the elements that he is calling a priori cannot be determined empirically, but despite the fact that they are capable of alteration, they are not arbitrary, given that the a priori concepts and principles are developed with the idea of applications to the empirical world in mind (Lewis 1929, 237–238). Lewis also gives a nod to the idea of a biological origin of some of our concepts, though this is not a permanent or fixed limitation on our thought. We can overcome what is natural to us and develop unnatural concepts in science if this is required to further our understanding and knowledge, according to Lewis:

> Some modes of thought are simpler and come more naturally to us than others which still are possible and which might, indeed, be called upon if an enlarged experience should sufficiently alter our problems—just as some modes of bodily translation are more easy and natural, though these may be somewhat altered when the environment includes a sufficient number of automobiles and airplanes. (1929, 238)

Lewis refers to Poincaré's "space and geometry" chapter of *Science and Hypothesis* in order to show that the place where the dividing line between space and matter is drawn is a pragmatic choice (1929, 253). In this chapter Poincaré sets out his distinction between geometric space and perceptual space and discusses the origin of geometry (1913, 75). Lewis seems to have in mind the idea that pure (metric) geometry is limited only by consistency and that the application of geometry to material bodies involves pragmatic choice.

Lewis sets his work in the context of the traditional debate between rationalism and empiricism. Rationalism is no longer credible because it claims that we have a priori knowledge through intuition or a "natural light." Changes in mathematics, especially the development of non-Euclidean geometries, and changes in natural science such as the theory of relativity have shown that what had formerly been taken to be known with certainty is not. On the other hand, the empiricist position cannot account for the a priori status of mathematics nor of the principles of physics. Lewis rejects an empiricist account of mathematics, siding with logicism, that is, he finds all of mathematics to be part of logic, and logic to be the prime example of the a priori. Thus, all of mathematics is analytic and indeed reducible to logic, but this move simply replaces the question of the status of mathematics with the question of the status of logic, which Lewis takes to be the prime example of a priori knowledge. Here things get subtle, given that although Lewis agrees with the Logical Positivist position that all a priori truths are analytic, his treatment of the question has a different emphasis. Like Reichenbach in his early work, Lewis accepts a constitutive role for a priori concepts and principles, and denies that the a priori is fixed or immutable.

While Lewis agrees with the Logical Positivists/Empiricists that all a priori truths are analytic, and that all synthetic truths are a posteriori, he usually does not put his point in these terms, except when he expresses this directly in appendix F to *Mind and World Order*, where he sketches the distinction between proposition in intension and in extension. "The first expresses in the predicate something logically contained in the subject; the subject-concept implies the predicate-concept. The second states a factual connection of two classes of objects" (Lewis 1929, 434). His point is that in the intension case, there is a necessary connection between the two concepts, whereas in the extension case, there is none, a distinction that relates to Lewis's long-standing complaint about material implication in logic. In the intension case, the implication is stronger, requiring something more that the mere fact that implication never leads from a true statement to a false one. Lewis seems to assume that we can tell unproblematically which properties are essential and which are not.

As I noted earlier, Lewis's pragmatic conception of the a priori refers to three conceptual areas, logic, sentences that are true by definition, and some elements of physical theories. The way that concepts apply to things in the world is what is most relevant to our discussion here, given that this is how a conventional element comes into play in our experience of the world. Revolutionary changes in science can appear as changes in fundamental concepts and the role of these concepts in knowledge is given by Lewis's pragmatic theory of the a priori. "Empirical knowledge arises through conceptual interpretation of the given" (Lewis 1929, 37). These conceptual interpretations are the focus of Lewis's theory of the a priori. It is important to note that Lewis sees his project as following directly from advances in mathematics and physics in the early twentieth century (Lewis 1929, vii–viii). Changes in mathematics and physics led to new philosophical thinking about the a priori and led to alternatives to the empiricist theory of the a priori.

PAP'S FUNCTIONAL A PRIORI

Arthur Pap shares a history with the Vienna Circle, although of course he was not a member. Like many members of the Vienna Circle, Pap was a European Jew whose family fled the Nazis, an immigrant scholar with a background in science as well as philosophy. He was younger, so his university education took place in the United States, first at Columbia, then at Yale for a master's degree under Ernst Cassirer, then back at Columbia for his PhD under Ernest Nagel, two more immigrant Jews who were interested in both science and philosophy. Cassirer was a neo-Kantian, Nagel was influenced by pragmatism, and each of these movements had a strong influence on Pap at the start of his remarkable career. He published a pair of articles in the *Journal of Philosophy* (1943a, 1943b) and one in *Philosophical Review* (1944) before completing his dissertation (1946), which won an award and was immediately published. In the course of his short career he published five books and more than fifty articles before dying of kidney disease at the age of thirty seven.

I am most interested here in Pap's work on what he called the functional a priori, which he develops in his dissertation, applying it especially to Newtonian physics. In Pap's theory of a functional a priori, he focusses on fundamental principles of science that are hardened into definitions and act as criteria for further inquiry. As I noted in Chapter 1, Friedman complains, with a certain amount of justice, that the fixing of formerly empirical statements is the same as Quine's hard core (Friedman 2001, 88n22). However, the principles that start out empirical and end up being fixed are only one type of constitutive element in science in Pap's view. Furthermore, the most important point about Pap's theory, as we will see, is that what is constitutive (or "functionally a priori") depends on context.

In developing this alternative theory of what had formerly been a priori knowledge, Pap was strongly influenced by the pragmatists C. I. Lewis ([1923] 1970, 1929) and John Dewey ([1938] 1986). Using Poincaré's conventionalism as a springboard, Pap attempted to substantiate his view with examples from physics, thus making his most significant foray into philosophy of science topics. Pap is of special interest in that while he was a fellow traveler and was writing in the heyday of Logical Empiricism, his philosophy of science took a pragmatic turn. As we will see, there is a direct line of descent from Pap to Lenzen, C. I. Lewis, Dewey, and from Dewey to Peirce.

Nonetheless, Pap did not remain long in the philosophy of science. After his dissertation, most of his work is more generally in analytic philosophy, not in the philosophy of science per se. He returned to philosophy of science only to write a textbook at the end of his career. Still, he is of interest to philosophers of science. Pap, and through him Lewis and Dewey, constituted an alternative philosophy of science in the 1950s that never quite took hold, despite the fact that their views on the a priori are very intriguing. Of course, there is more than this one version of the Vienna Circle–inspired Logical Empiricism. George Reisch has shown how Philip Frank and Otto Neurath, the left wing of the Vienna Circle, presented another alternative philosophy of science to that established in the 1950s (Reisch 2005). Dewey's *Logic* ([1938] 1986) would be the best candidate for such an alternative, but for whatever reason, it was not picked up by the philosophy of science community. On one hand, it was very positively reviewed at the time of its appearance, but it was also criticized savagely by Bertrand Russell. It is also true that Dewey was not up to speed on modern logic or modern physics, so his *Logic* could have looked rather less up to date and exciting when compared to that of the Logical Empiricists.

Pap's debt to pragmatism goes beyond the influence that C. I. Lewis had on him with his pragmatic theory of the a priori. Pap also drew directly on Dewey, as we can see in his numerous citations of Dewey and in his use of a phrase taken from Dewey that is important in making the connection between temporal change and the status of fundamental principles in science. In his *Logic*, Dewey says of fundamental principles:

> While they are derived from examination of methods previously used in their connection with the kind of conclusion they have produced, they are *operationally a priori* with respect to further inquiry. (Dewey [1938] 1986, 21, emphasis in the original)

In a footnote, Dewey calls his assertion "a free rendering of Peirce" and cites two passages in Peirce's collected papers (1958–1960, III:para. 154–168; V:para. 365–370). Pap quotes part of this phrase from Dewey without citation in his 1944 article "The Different Kinds of A Priori" (2006, 72), as if he were quoting from memory. In his dissertation, Pap again quotes Dewey without citation, but this time he uses a slightly different phraseology: "Hooke's law remains refutable by experience, even though it may provisionally function as 'a priori with respect to further operations,' in Dewey's phrase" (Pap 1946, 30). Since Pap quotes Dewey without a citation, it is hard to know whether he has just misremembered Dewey's phrase that he quoted earlier, or if he has some other passage in Dewey in mind. Given that the *Logic* is the only work of Dewey's that is cited in Pap's bibliography in his dissertation and that Dewey's phrase only exists in the *Logic* as quoted earlier, it seems likely that Pap has given this quotation from memory and that he has not gotten Dewey's phrase exactly correct. The lack of citations in the two passages is unusual for Pap and could represent a special familiarity with Dewey's work, as if Dewey were on his mind while writing. The phrase and the idea that it represents obviously impressed Pap, given that he quotes it in two different works.

In Dewey as in Pap, what starts out as empirical knowledge can end up being fixed and taken for granted, that is, it can function as a priori knowledge, or as I would prefer to say, it can be constitutive. Most important to note is the temporal aspect of this conception—that is the point of the "with respect to further operations (or inquiry)" language. A principle that started out as provisional and empirical will be treated as a priori at some other point in time or in some other context. On this view everything is ultimately provisional, but some elements of our knowledge must be taken as fixed at a given point in order to pursue further inquiry. There is no mention of there being particular kinds of principles that are likely to function as a priori elements. As Pap emphasizes in the introduction to his work, the temporal aspect of functionally a priori knowledge is the crucial point that allowed him to see his way out of his discomfort over the positivists' treatment of the a priori (Pap 1946, vii). The same statement can play a very different role in a scientific theory at different times. I discuss this later when I present Pap's dissertation, but here I would like to also note that mutatis mutandis, the same temporal solution can be applied to the problem of distinguishing between analytic and synthetic statements, given that what was once synthetic can be taken to be analytic, and what was once analytic, in being questioned, can be taken as synthetic and empirical, at a different time and in a different context.

Pap explicitly acknowledges the lineage of influences that permeates his dissertation. He is working with a particular interpretation of Poincaré, that given by Victor Lenzen (1931), and with Lewis's and Dewey's pragmatic conceptions of the a priori, which leads him to view functionally a priori principles as empirical laws that have been taken to be definitions and function as categorical principles or constitutive conditions:

> As the *motto* which is prefixed to this essay indicates, Poincaré's "conventions" (in mechanics) are principles that have been "derived" from experimental laws—in much the same way as our "real definitions" are thus derived—at the same time, however, [they] have been immunized against possible invalidation by future experience. They function like Lewis' "categorial principles," in so far as, if an experience does not fit them, "so much the worse for the experience." Our emphasis on this conventionalization of inductive generalisations has been greatly influenced by Victor Lenzen's detailed application of Poincaré's notion of "conventions" to physical theory. (Pap 1946, 21)

It is important to note that the conventionalism mentioned here is that of the principles of mechanics, not Poincaré's geometric conventionalism, which has a different source. Lenzen has a pragmatic interpretation of Poincaré's conventionalism of principles, an interpretation that has a lot to say for itself. Poincaré himself was not influenced by pragmatism, but his work does show some similar themes, such as the human contribution to knowledge exhibited in his conventionalism. Rene Berthelot was the first to notice the connection between Poincaré and pragmatism and his study is groundbreaking in this regard (Berthelot 1911, vol. 1).

Neo-Kantianism is important because it formed a context for the work of the Pragmatists, the Vienna Circle and to Poincaré's work. It grounds the focus on the human element of knowledge and provides the basic distinctions between the a priori and empirical and the analytic and synthetic. Neo-Kantianism was also a direct influence on Pap's functional theory of the a priori. Indeed, Pap, in "On the Meaning of Necessity" (1943), claims that Cassirer's interpretation of Kant is a version of his own functional theory of the a priori: "Cassirer tends to assimilate Kant's doctrine of the a priori to the functional-pragmatic interpretation of the a priori" (2006, 53). So, Pap never sees any conflict between the neo-Kantians and the pragmatists on the functional account of the a priori. He relates Lenzen's views to Kant in the following passage, which also underscores why I argue that we should give up the term *a priori*:

> Lenzen is Kantian in so far as he acknowledges that universal principles enter as essential determinants into what the physicist declares as "reality." These "constitutive conditions of experience," however, are, for Lenzen, "synthetic a priori" only in the crooked sense of being, on the one hand, empirically grounded, and on the other hand, a priori in their "constructive function." (Pap 1946, 21)

That is, universal principles are synthetic in so far as they are empirically grounded—they are statements about objects in the real world—but they are a priori only in a functional sense, given that they are neither certain nor known through a special intuition, thus sharing almost none of the characteristics of what has classically been called a priori knowledge. Contrary to the view of the Logical Empiricists, however, they are not simply analyzable in terms of meaning as analytic statements. Pap and Lenzen's theory of the functional a priori is close to Reichenbach early view, expressed in *The Theory Relativity and A Priori Knowledge* (1920), which was discussed in Chapter 4.

PAP'S EARLY ARTICLES

Prior to writing his dissertation, Pap published two articles in the *Journal of Philosophy* and one in *Philosophical Review*. The most relevant one for my purposes is the last one, called "The Different Kinds of A Priori" (1944). In this article, Pap distinguishes three kinds of a priori, advocating the view that all three kinds of a priori knowledge exist. He calls the first the formal or analytic a priori and it includes all tautologies and other analytic truths. Next is the functional a priori, statements that play the role of a priori knowledge but are hypothetical or conditional, that is, they are "predicable of conceptual means in relation to objectives or ends of inquiry" (2006, 57). Finally, he advocates for what he terms a material a priori, that is, a synthetic a priori that is supposedly self-evident. He says that he avoids using the term 'synthetic a priori' because some of Kant's synthetic a priori is actually the functional a priori, that is, constitutive but changing status in different contexts. Pap's main example of the material a priori are the principles of logic themselves, which is a little odd, given that tautologies and other logical principles are supposed to be ipso facto analytic. Could we not define the principles of logic as analytic? In any case, Pap's point is that the principles of logic have a special status and he is surely right about that, since they must be presupposed before any inquiry into the analytic/synthetic distinction can be started:

> Thus the very analysis of what is meant by the "formal *a priori*" reveals the existence of another kind of *a priori* without which the formal

a priori could not even be defined. I shall call this the *material a priori,* avoiding the more familiar term "synthetic," because the latter has been ambiguously applied, by Kant, to both the material and the functional *a priori*. The principles of logic themselves, which we just saw to be essentially involved in the definition of the formal *a priori,* are materially *a priori*. ([1944] 2006, 63–64)

Pap's insistence on the material a priori in this early article puts him at odds with both Logical Empiricism and with C. I. Lewis. He acknowledges this in rather tongue-in-cheek fashion, noting that he could be accused of mystical Platonism. Rather than forthrightly embracing such mysticism, however, Pap offers a way out, an alternative to seeing logic either as given by pure intuition or as taken it to be merely conventional. Speaking of the laws of logic, Pap says,

> Their truth is a matter neither of deduction nor of induction. We are then left with two alternatives: either they are self-evident, "seen" to be true in pure "Wesensanschauung," as the phenomenologists would say, or they are conventions. Even though it would certainly be more emancipated to accept the latter alternative, and to dismiss the former alternative as mystical Platonism, I venture to suggest that we are not, here, really confronted with mutually *exclusive* alternatives. Just as empirical laws of nature are used as conventional definitions of empirical concepts, because they are *true* in a non-pragmatic sense, so the principles of logic can be used as implicit definitions of logical concepts, like negation, implication etc., because they possess some kind of evidence that is independent of the use that can be made of them. ([1944] 2006, 63–64)

Pap's idea here is that the principles of logic can function as first principles no matter what their epistemological status. The fact that they can be used to define the logical connectives does not show that they are "merely conventional." Even if we grant this point to Pap, however, it still leaves the self-evident character of logic to be explained. The point of the argument for the conventionalism of logic is precisely to do away with this question— that is, to get rid of the demand for an explanation of the epistemological ground of the principles of logic. It seems that Pap is still left with some commitment to a form of self-evidence, even if he is not willing to embrace it fully. In his later writings he is likewise circumspect about the status of self-evident principles. We can only note here that Pap is committed to them.

Since I am following Pap's idea of the functional a priori here, his discussion of this type of a priori knowledge is most relevant. However, Pap does no more than introduce the idea in his 1944 article, given that the majority of the article is devoted to the formal and the material a priori. Still, it is important to note that Pap developed the idea of the functional a priori before writing his dissertation. The main idea of the functional a priori is that one considers the role that a sentence is playing in a physical theory, rather than its origin. Certain ideas become criteria for the existence of phenomena, becoming definitions if they are used as criteria for the establishment of a correct labeling of phenomena in a category. In his 1944 article Pap sets out the functional a priori as follows:

> Any synthetic sentence, whether empirical or "eidetic," may be *made,* by a conventional act, into an analytic, formally *a priori* sentence. But it is usually *made* formally *a priori,* in order to be *taken* as *functionally a priori,* i.e., a hypothetically necessary presupposition, a "procedural means," as Dewey would say; or, as Kant called these functionally *a priori* principles, a "Grundsatz," in contradistinction to an analytically demonstrable "Lehrsatz." ([1944] 2006, 70 emphasis in the original)

These principles are hypothetically necessary because they are the premises on which theories are founded. Their origin, whether empirical or the result of some sort of rational intuition, does not diminish their role as an analytic sentence in a physical theory. Thus, Pap focuses on the role that a sentence plays in a physical theory, rather than on the origin of the sentence.

It is important to note that even in this early article, Pap emphasizes that the three types of a priori sentences he distinguishes do not become a rigid trichotomy. Sentences can play different roles at different times:

> In summary, it should be noted that, although the formal *a priori,* the material *a priori,* and the functional *a priori* are, as categories or epistemological predicates, *distinct,* they are quite compatible in the sense of being predicable of one and the same sentence. As a matter of fact, the main intent of this analysis has been to mark out the different types of epistemological status that accrue to statements in the process of scientific systematization. Synthetic statements, whether they are empirical or materially *a priori,* are *made* into analytic statements in order to be *taken* as "leading principles" or "conventions." Hypostatization of the categorial *distinction* between synthetic truth and conventional-analytic definition into existential *separation,* such as to think of the statements which are epistemologically qualified by these categories, as of mutually exclusive classes, gives rise to a radical misconception of science. (Pap [1944] 2006, 74, emphasis in the original)

What Pap seems to have in mind is the kind of process that he thinks takes place in the functional a priori. Sentences that have an empirical origin can be taken to be analytic if we make a decision to take the sentence as a principle that no empirical evidence can directly overturn. Therefore, there is certainly no permanent classification of sentences; they can change from empirical to analytic. He seems to mean something further, however, namely that there is no strict distinction between the different types of a priori, leaving borderline cases that are hard to classify. This would presumably mean that some logical principles could seem to be merely analytic rather than materially a priori and vice-versa, namely, that some analytic truths are close to being logical principles.

DISSERTATION: *THE A PRIORI IN PHYSICAL THEORY*

Pap opens his discussion of the a priori in physical theory with the following quote from Poincaré's *Science and Hypothesis*, which was mentioned earlier as his motto:

> Principles are conventions and disguised definitions. They are nevertheless drawn from experimental laws, these laws have been so to speak erected as principles to which our mind assigns an absolute value. (Poincaré 1913, 125, translation modified. Pap quotes the original French)

He then lists his other influences and states the basic insight that led him to his functional theory of the a priori.

> Under the influence, first, of C.I. Lewis' "conceptual pragmatism," as developed in *Mind and the World Order*, and then of Dewey's *Theory of Inquiry*, I was led to develop a functional interpretation of the a priori with close regard to the methods of physics. (Pap 1946, vii)

It is interesting that Pap refers to Dewey's *Logic: The Theory of Inquiry* by its subtitle. Although Pap is informal in other citations of Dewey, as we have seen earlier, Pap's emphasis on the theory of inquiry is correct, given the focus of Dewey's book. The subtitle more accurately reflects the content of the book than the main title, especially in a context in which symbolic logic is ascendant. Coming as it does, after *Principia Mathematica* and the rise of symbolic logic, Dewey's title seems incongruous. Of course, Dewey is standing in a long line of succession of books on logic that we would now think of as epistemology, with Dewey's *Logic* being perhaps the last major work in this tradition. Dewey explicitly and repeatedly refers to Aristotle's and Mill's works on logic as his precursors.

Pap sets out the problem that he will discuss in his dissertation by telling why he rejects the Logical Empiricist position on a priori knowledge:

> The dictum that in so far as a statement is a priori it is verbal and "asserts nothing about reality" and in so far as it is synthetic it may be refuted at any moment by experience, always left me with a sense of mental discomfort. After several attempts at rehabilitating the honorable status of "synthetic a priori" propositions had failed, the conventionalist writings of Duhem and Poincaré, and especially Victor Lenzen's *The Nature of Physical Theory*, helped me to locate the trouble. If, as methodologists, we adopt a static point of view, and examine the body

of scientific propositions as it may be found systematized at a definite stage of inquiry, we will, indeed, successfully divide it into analytic and synthetic propositions, as forming mutually exclusive classes. If, however, our point of view is dynamic or developmental, we shall find that what were experimental laws at one stage come to function, in virtue of extensive confirmation by experience, as analytical rules or "conventions," in Poincaré's language, at a later stage. (1946, vii)

Thus, Pap's key insight is that the status of scientific statements can change over time and from one context to another. The result is that while there is no a priori in the traditional sense, there still are and must be scientific statements that play the role of principles and function as a priori knowledge, even if they were originally empirical.

As Pap notes, this view has its roots in Poincaré's conventionalism. Unfortunately, Poincaré's conventionalism has been open to many interpretations and controversies, so saying that Pap's theory of the a priori has Poincaré's conventionalism as a source may add more confusion and controversy than it adds to a clarification of Pap's view. As I argued in Chapter 3, there is an important distinction between the motivation for Poincaré's geometric conventionalism and his conventionalism about fundamental principles of science. Poincaré's conventionalism of principles fits the reading given by Lenzen and Pap, while the geometric conventionalism does not. Poincaré thinks that some empirical laws of science can be "erected" as principles that have a functionally a priori status, but geometry has a special status as being neither empirical nor a priori.

My main point here is that we can see Pap as working in a tradition, one that constitutes an alternative to Logical Empiricist orthodoxy, even if Pap is writing at the time of the heyday of Logical Empiricism and is very connected to it. On one hand, the questions that are being posed and the distinctions that are being used are very much the same as those of the Logical Empiricists. Both share a neo-Kantian background that forms the context to these debates. On the other hand, Pap does not accept the complete elimination of a priori knowledge, as do the Logical Empiricists.

As some of his contemporary critics noted, one may wonder why Pap wants to hold on to the notion of a priori knowledge when, in many ways, the functional a priori is not a priori at all in the traditional sense. I have suggested that it would be better to say that Pap and the pragmatists have a theory of the constitutive elements in science and to drop the usage of 'a priori' altogether. I think that it is the neo-Kantian heritage that makes Pap hold onto this language. Pap does not want to break totally with Kant by rejecting the a priori outright, given that the constitutive role of the fundamental principles of scientific theories is too important to gloss over (1946, viii). Pap agrees with Dewey that the functional a priori is contextual, showing that his view has some elements of Carnap's mature position. Statements are analytic in a language for Carnap—statements are a priori or empirical in a context, according to Pap. What Pap calls the functional a priori elements in science are what I am calling the constitutive elements.

THE ANALYTIC FUNCTIONING OF EMPIRICAL LAWS

In his discussion of the "hardening" empirical generalizations to definitions, Pap opens with the distinction between real and nominal definitions. Real definitions are based on facts in the world, while nominal definitions are purely verbal. He then endorses Hilbert's notion of implicit definition of the primitive terms of geometry and extends this to cases in natural science, specifically Newtonian mechanics. The central claim of Pap's functional theory of the a priori is that empirical (and hence contingent) statements may sometimes function as temporarily fixed principles. They are taken for granted as true and used as guideposts by which phenomena are interpreted, but they can be changed later:

> As inductive generalisations become increasingly confirmed they tend to be used as principles by which the "phenomena" are interpreted. For example, within Newtonian dynamics it would hardly ever occur to a physicist to explain the negative outcome of an astronomical prediction in terms of a failure of the general equation of motion; assuming the latter to be valid, the discrepancy between observation and prediction will "prove" to him that something is wrong with his assumptions concerning the initial and boundary conditions. (Pap 1946, 28)

The perturbations in planetary motion might be explained by predicting the existence of an unknown planet that is causing the changes in orbit, rather than inferring that there is something wrong with the fundamental laws of motion. If a body is moving at a velocity other than $\frac{1}{2}gt^2$, we will assume that it is not in free fall and that forces other than gravity are acting on it. Why are such principles never called into question? Logically speaking they can be questioned, of course, but at a given developmental stage of a science, some principles are simply taken to be true. They are no longer in question or up for grabs; rather, they are used to make predictions and to interpret phenomena.

Pap says that he will emphasize the making and functioning of a priori sentences in science, rather than focus on the ultimate distinction between a priori and empirical statements, as does Lewis (Pap 1946, 4). Lewis takes the a priori to be analytic, but Pap is closer to the truth when he says that a priori statements are empirical and then hardened into analytic statements, which is why the distinction between analytic and synthetic statements is much less clear cut that Lewis would have us believe. Pap's recognition that there is no clear-cut distinction between analytic and synthetic statements is more defensible than Lewis's dichotomy.

Pap also criticizes Lewis for characterizing the a priori in terms of what one is willing to maintain in the face of all experience (Pap 1946, 4; see Lewis [1929] 1970, 224). Since Duhem has shown that it is always logically possible to hold onto a given set of ideas and to explain negative experimental results by adjusting auxiliary hypotheses, every statement could be considered a priori by this criterion, if we mean simply that it is possible to maintain the proposition.

But Lewis would not seem to have such a weak criterion in mind. What we are willing to maintain is surely narrower than what is logically possible. So Pap argues that if Lewis means that it must be practically possible to hold onto a given set of ideas for them to be a priori, then it would become extremely complicated to decide whether a given statement is a priori (or analytic) or not. Pap is right, but I think that he cannot use this point to criticize Lewis. We have to own up to the fact that it is extremely difficult to decide what is a priori and what is empirical. Just as it is impossible to make a sharp divide between analytic and synthetic statements, it is also impossible to make a sharp divide between a priori and empirical statements. Indeed, Pap focuses on those empirical statements that are hardened into definitions, that is, those that are empirical at one time or in one context, and function as a priori in another. So Pap is right, but he cannot use the point to criticize Lewis quite as he imagines. The problem that should be in focus is that Lewis thinks that it is possible to make an absolute distinction between synthetic and analytic statements, while in fact, the distinction can only be made for a certain time and in a certain context. What Pap correctly proposes is to look at the function that the sentence plays in a given physical theory, in order to determine if the sentence is a priori or not. Pap emphasizes the process by which inductive generalizations can be taken to be definitions and hence become analytic statements. They then become a priori conventions by which scientific theory is constituted.

In a final criticism of Lewis, Pap tries to argue that there is a conflict between Lewis's claim that the a priori is analytic and his thoroughgoing fallibilism:

> On the one hand, Lewis insists on the analytic or definitive nature of these "criteria of reality." On the other hand, he claims that, since our classificatory judgments imply an indefinite number of other classificatory judgments in terms of which they may be verified, complete verification of our classifications is impossible and hence any subsumptive judgment is a merely probable hypothesis. But these two emphases are hardly compatible. (1946, 3)

The problem is that meaning has an empirical basis and that, as such, it is contingent. It is not clear to me that this is a real problem for Lewis, however. The solution to the apparent contradiction is that Lewis's criterion of reality needs to be taken as contingent as well. Pragmatists should be expected to take everything to be contingent and revisable, including definitions.

LATER WORKS

One postdissertation work of Pap's that is worth considering in this survey of his philosophy of science is "Does Science Have Metaphysical Presuppositions?" which is chapter 16 of *Elements of Analytic Philosophy* ([1949] 1972) and was reprinted in slightly abridged form in Feigl and Brodbeck's hugely influential collection *Readings in the Philosophy of Science* (1953, 21–33). Pap's main argument in this piece is that while science has presuppositions, these should not be called metaphysical. Specifically, he says that while the presuppositions are statements of great generality, there is nothing to be gained by calling them metaphysical. This line of argument seems to me to sidestep the interesting questions, namely, what kinds of presuppositions are required in science and what do we say about their status. Pap is surely right that the label 'metaphysics' is not important, but the status that presuppositions have as what we might call pre-empirical is highly important.

The most interesting section of the chapter in that regard refers to justification of presuppositions. Pap argues that they are ultimately empirical, even if we have forgotten the empirical origin of the presupposition, but that they come to be treated as a priori:

> it happens all the time that a principle which has been highly confirmed in past inquiries comes to function as a practically unquestioned assumption in future inquiries; and if we forget the empirical background of the principle we might be tempted to regard it as a necessary or *a priori* truth, a genuinely metaphysical presupposition of scientific procedures. ([1949] 1972, 440)

It is not really the case that these principles were learned empirically first and that then the origin of the knowledge was forgotten, or at least not in all cases. Sometimes a principle comes from a former scientific dispute long since resolved, as in the case of spontaneous generation for example, but sometimes it seems that the empirical origin of the principle is questionable, as in the case of Newton's first law, the principle of inertia, which is discussed in Chapter 6.

POINCARÉ'S GEOMETRIC CONVENTIONALISM

In his textbook on the philosophy of science, Pap (1962) clearly follows Reichenbach's interpretation of geometric conventionalism. Pure geometry is purely formal and neither true nor false, while applied geometry is conventional in the sense that we can choose a pure geometry plus a set of coordinating definitions in order to create a physical theory. Once we have chosen a set of coordinating definitions, there is a truth of the matter of which geometry applies to physical space:

> the question is: "is it a fact that the Riemannian postulates are true of physical space, and not the Euclidean postulates?" the proper answer is not so much "no" than: the question is meaningless. The postulates of a pure geometry are no statements at all, hence there is no sense in saying that they are true or false of anything. If there is an element of *convention* in the application of a geometry to physical space, as Poincaré

emphasized, it is this: if a given set of coordinative definitions converts the postulates of a given formal geometry into empirically false statements, there are two alternative ways of amending the situation, and it is a matter of convenience which may be chosen: one might change the coordinative definitions and stick to the use of just that formal geometry; or one might stick to the coordinative definitions and look for another formal geometry which the very same coordinative definitions would convert into a true physical geometry. (Pap [1949] 1972, 428)

As I argued in Chapter 3, this is not all that is going on in Poincaré's conventionalism. If it were, Poincaré's would have to be a thoroughgoing conventionalist, not merely a conventionalist about geometry and a few principles but about the rest of physical theory as well, which is not the case, as Poincaré makes abundantly clear, especially when distancing himself from the views of LeRoy. In fact, Poincaré's conventionalism about geometry and his conventionalism about principles are different, the first being based on a relational theory of space and the second on the functional theory of the a priori. Poincaré explicitly distinguishes conventions in geometry from the conventions of principles in the physical sciences (1913, 124). In adopting Reichenbach's interpretation of Poincaré's geometrical conventionalism in his later writings, Pap extends his theory of the functional a priori too far. The conventionalism of principles should be the only focus of Pap's investigation of the functional a priori.

MEANING AND MATHEMATICS

In the end, Pap does not want to say, it appears, that the synthetic a priori is real and that it is impossible to analyze all a priori statements as analytic. He takes a more ambiguous line, arguing for the functional a priori while maintaining an empirical stance. Pap is somewhat misled, I believe, by his focus on meaning, that is, his shift from the philosophy of science proper to general methods of analytic philosophy. For example, Pap insists that empirical sentences that have been hardened into principles have undergone a shift of the meaning of their terms:

> Our contention that scientific truths undergo a development from contingency to analytic necessity (in Poincaré's language, experimental laws are "erected" into "conventions") might then be reformulated as follows: one and the same sentence which is in one context of inquiry synthetic may, in a different context of inquiry, be analytic, in virtue of a shift of *meaning* undergone by some of its terms. (1946, 24, emphasis in the original)

Presumably, what is at stake here is that the meaning of a term has changed to include the property that is in question when the sentence becomes a criterion. So, falling at the rate of $\frac{1}{2}gt^2$ becomes a criterion of free fall when we no longer consider that the law of free fall might need correction and, instead, take any other measured rate of free fall as evidence for the existence of force acting on the body. Pap would have us then include ½gt² in the definition of *free fall*, but I do not see what is gained here by talking in terms of meaning. Why not simply say that we now have a criterion of free fall and leave our theory neutral when it comes to questions of meaning? I am suspicious of discussions of meaning, because I think they play little role in science as it is practiced. Of course, terms are defined in science, but discussion of a theory of meaning belongs to linguistics and philosophy. Couching the discussion of the functional a priori in terms of meaning is not helpful and if anything, just leads to confusion about the functional status of the principles of science, which function as a priori elements, that is, as taken-for-granted principles.

As noted previously, Pap recognizes that the distinction between analytic and synthetic statements cannot be fixed permanently and in advance of inquiry:

> With regard to existential inquiry, analyticity is an ideal limit. The concepts 'analytic' and 'synthetic' are no doubt mutually exclusive, but if scientific inquiry is viewed as a continuous process, it will be seen that empirical laws converge from the status of synthetic descriptions towards the status of analytic "criteria of reality" (it being understood, though, that "criteria of reality" have no *final* status but are *instrumental* in further inquiry). (1946, 26, emphasis in the original)

I want to suggest that thinking that the analytic and the synthetic can be determined prior to and independent of empirical inquiry is a fundamental mistake made by both Lewis and Carnap. Both want to separate the formation of a language from the development of physical theory, but the course of empirical theory can itself influence what is taken to be analytic or fixed and what is taken to be synthetic or empirical. It is impossible to set out the meaning of terms independently of physical theory.

The role of mathematics in physical theory is a case in point. Mathematical statements did not start out as empirical and become hardened into principles that have the status of a priori knowledge. I do not attempt to answer the question of what the epistemological status of mathematics is, but rather argue that the functional theory of a priori knowledge is neutral with regard to theories of the status of mathematics. Concerning logic and mathematics, Pap advocates the continuity of mathematical and physical inquiry:

> We want, however, to emphasize that, in respect of this analytic functioning, there is but a difference of degree between mathematical or logical truths on the one hand, and highly warranted inductive generalisations, on the other hand. It is true that the failure of a physical prediction would never be explained by casting suspicion upon the logical and mathematical principles that are implicitly assumed in the deduction of testable

consequences. But the same holds true, though to a less degree, with respect to fundamental physical principles such as the general equation of motion or the conservation principles. If, therefore, the validity of logico-mathematical truths is to be *sui generis,* their apriority cannot be defined in terms of their prescriptive function in empirical inquiry. (1946, 28)

This is not quite to say that mathematics is empirical but, rather, that the function of mathematics in physical theory is very similar to that of fundamental principles that are empirical in origin. Mathematical truths are found by conceptual practices and 'hardened' when they are taken up in physical theory where they are taken for granted as true in the new context. The ultimate ground of pure mathematics, whether it be logic, intuition, or syntax, is not important in this context. The functionally a priori role of mathematics in physical theory is consistent with all three of the classic standpoints in the philosophy of mathematics and there is no reason that it would not be compatible with others as well. The important point here is the role that mathematics plays in physical theory, not the ultimate status of mathematics.

CONCLUSION

The epistemological status of logic, mathematics and even some fundamental laws of nature have always been problematic. They do not seem to be empirical or to be simply a matter of definition. However, taking them to be known a priori is not viable either, given that there is no intuitive, certain knowledge of nature. The pragmatic or functional theory of the former a priori addresses this dilemma by focusing on the special epistemological status that fundamental laws of nature and parts of mathematics have as constitutive parts of physical theory. These elements of physical theory must be assumed before it is even possible to begin inquiry into a given science. This constitutive function gives principles of physics and parts of mathematics their distinctive epistemological status.

I have presented an overview of Lewis's pragmatic a priori and Pap's functional theory of a priori knowledge and shown them to be grounded in pragmatism and neo-Kantianism and to be a solution to the problem of the epistemic status of mathematics and certain fundamental principles of physical theory. They are two kinds of theories of the constitutive elements in science. The pragmatic conception of the a priori, developed by Dewey and Lewis, and the function theory of the a priori, developed by Pap, present alternatives to the Logical Empiricist conception of a priori knowledge and thereby to the dominate philosophy of science in the twentieth century. These alternatives have largely been forgotten, and even though they were discussed in their time, they never fully took hold in the philosophy of science. I hope to contribute to reviving a discussion of them now.

NOTES

1. For discussion of the neo-Kantian treatment of science as a given, see Richardson (2006).
2. Some have read Cassirer as holding a form of structural realism in which the functional elements are taken as fundamental. See Slowik (2012) for an overview.
3. See Friedman (2012, 47–48) for discussion of the distinction between regulative and constitutive and the issues that this distinction raises.
4. Hasok Chang (2008) and Thomas Mormann (2012a, 2012b) have both compared Lewis to Friedman. Chang also makes some interesting comparisons between Lewis and Friedman on the continuity of scientific theory, a topic I take up in Chapter 7.
5. Later reprints of Einstein's book have different pagination. I refer to the edition used by Lewis.

BIBLIOGRAPHY

Berthelot, René. 1911. *Un Romantisme Utilitaire: Étude Sur le Mouvement Pragmatiste: le Pragmatisme chez Nietzsche et chez Poincaré.* Paris: Félix Alcan.
Cassirer, Ernst. 1923. *Substance and Function* (1910) and *Einstein's Theory of Relativity* (1921). Chicago: Open Court.
———. 1987. *Zur Modernen Physik.* Darmstadt, Germany: Wissenschaftliche Buchgesellschaft.
———. 2009. *Nachgelassene Manuskripte und Texte.* In *Ausgewählter wissenschaftlicher Briefwechsel*, vol. 18, edited by J. M. Krois. Hamburg: Meiner.
Chang, Hasok. 2008. "Contingent Transcendental Arguments for Metaphysical Principles." In *Kant and Philosophy of Science Today*, edited by M. Massimi. Cambridge: Cambridge University Press, 113–133.
Dewey, John. (1938) 1986. *Logic: The Theory of Inquiry.* New York: Holt, Rinehart and Winston. Rpt. *Collected Works of John Dewey: The Later Works.* Vol. 12. Carbondale: Southern Illinois University Press.
Einstein, Albert 1920 *Relativity: The Special and General Theory.* Translated by R. W. Lawson. New York: Henry Holt and Company. Original German edition 1916, translated from the third German edition, 1918.
Feigl, Herbert, and May Brodbeck, eds. 1953. *Readings in the Philosophy of Science.* New York: Appleton-Century-Crofts.
Ferrari, Massimo. 2009. "Is Cassirer a Neo-Kantian Methodologically Speaking?" In *Neo-Kantianism in Contemporary Philosophy*, Edited by R. A. Makkreel and S. Luft. Bloomington: Indiana University Press, 293–312.
———. 2012. "Between Cassirer and Kuhn. Some Remarks on Friedman's Relativized A Priori." *Studies in History and Philosophy of Science Part A* 43 (1): 18–26.
Friedman, Michael. 2000. *A Parting of the Ways: Carnap, Cassirer, and Heidegger.* Chicago: Open Court Press.
———. 2001. *Dynamics of Reason: The 1999 Kant Lectures at Stanford University.* Stanford, CA: CSLI Publications.
———. 2011. "Ernst Cassirer." *The Stanford Encyclopedia of Philosophy* (Spring 2011 Edition), edited by E. N. Zalta. http://plato.stanford.edu/archives/spr2011/entries/cassirer/.
———. 2012. "Reconsidering the dynamics of reason: Response to Ferrari, Mormann, Nordmann, and Uebel." *Studies in History and Philosophy of Science Part A* 43 (1): 47–53.

Heis, Jeremy. 2014a. "Realism, Functions, and the a priori: Ernst Cassirer's Philosophy of Science." *Studies In History and Philosophy of Science Part A* 48 (0): 10–19.

———. 2014b. "Ernst Cassirer's Substanzbegriff und Funktionsbegriff." *HOPOS: The Journal of the International Society for the History of Philosophy of Science* 4 (2): 241–270.

Janis, Allen. 2014 "Conventionality of Simultaneity." *The Stanford Encyclopedia of Philosophy* (Fall 2014 Edition), edited by Edward N. Zalta. http://plato.stanford. edu/archives/fall2014/entries/spacetime-convensimul/.

Lenzen, Victor. 1931. *The Nature of Physical Theory: A Study in Theory of Knowledge*. New York: John Wiley and Sons.

Lewis, Clarence Irving. (1923) 1970. "A Pragmatic Conception of the A Priori." In *Collected Papers of Clarence Irving Lewis*, edited by J.D. Goheen and J. John L. Mothershead. Stanford, CA: Stanford University Press, 231–239.

———. (1929) 1956. *Mind and The World Order: an Outline of a Theory of Knowledge*. New York: Charles Scribner's Sons. Rpt. New York: Dover.

Lovrenov, Maja. 2006. "The Role of Invariance in Cassirer's Interpretation of the Theory of Relativity" *Synthesis Philosophica* 42 (2): 233–241.

Mormann, Thomas. 2012a. "A place for pragmatism in the dynamics of reason?" *Studies in History and Philosophy of Science Part A* 43 (1): 27–37.

———. 2012b. "Toward a Theory of the Pragmatic A Priori: From Carnap to Lewis and Beyond." In *Rudolf Carnap and the Legacy of Logical Empiricism*, edited by R. Creath. Dordrecht: Springer, 113–132.

Norton, John D. 2014. "Einstein's Special Theory of Relativity and the Problems in the Electrodynamics of Moving Bodies that Led Him to it." In *Cambridge Companion to Einstein*, edited by M. Janssen and C. Lehner. Cambridge: Cambridge University Press, 72–102.

Pap, Arthur. 1943a. "On the Meaning of Necessity." *Journal of Philosophy* 40 (17): 449–458. Reprinted in Pap 2006, 47–56.

———. 1943b. "On the Meaning of Universality." *Philosophy* 40 (19): 505–514.

———. 1944. "The Different Kinds of A Priori." *Philosophical Review* 13 (5): 465–484. Reprinted in Pap 2006, 57–76.

———. 1946. *The A Priori in Physical Theory*. New York: King's Crown Press.

———. [1949] 1972. *Elements of Analytic Philosophy*. New York: Macmillan. Rpt. New York: Hafner.

———. 1962. *An Introduction to the Philosophy of Science*. New York: Macmillan.

———. 2006. *The Limits of Logical Empiricism: Selected Papers of Arthur Pap*. Edited by Alfons Keupink and Sanford Shieh. Dordrecht: Springer.

Peirce, Charles S. 1958–1960. *Collected papers*. 8 vols. Edited by Charles Hartshorne and Paul Weiss. Cambridge, MA: Belknap Press of Harvard University Press.

Poincaré, Henri. 1913. *The Foundations of Science: Science and Hypothesis, the Value of Science, Science and Method*. New York: The Science Press.

Quine, Willard Van Orman. 1953. "Two Dogmas of Empiricism." In *From a Logical Point of View*. Cambridge, MA: Harvard University Press, 20–46.

Reichenbach, Hans. (1928) 1958. *Philosophy of Space and Time*. New York: Dover.

Reisch, George A. 2005. *How the Cold War Transformed Philosophy of Science to the Icy Slopes of Logic*. Cambridge and New York: Cambridge University Press.

Richardson, Alan. 1998. *Carnap's Construction of the World: The Aufbau and the Emergence of Logical Empiricism*. Cambridge: Cambridge University Press.

———. 2006. "'The Fact of Science' and Critique of Knowledge: Exact Science as Problem and Resource in Marburg Neo-Kantianism." In *The Kantian Legacy in Nineteenth Century Science*, edited by M. Friedman and A. Nordmann. Cambridge, MA: MIT Press, 211–226.

Ryckman, Thomas. 2003. "Two Roads form Kant: Cassirer, Reichenbach, and General Relativity." In *Logical Empiricism: Historical and Contemporary*

Perspectives, edited by P. Parrini, W. C. Salmon, and M. H. Salmon. Pittsburgh, PA: University of Pittsburgh Press, 159–193.

———. 2005. *The Reign of Relativity: Philosophy in Physics, 1915–1925*. New York: Oxford University Press.

Schlick, Moritz. 1921. "Critical or Empiricist Interpretation of Modern Physics? Remarks on Ernst Cassirer's Einstein's Theory of Relativity." In *Philosophical Papers*, vol. I (1909–1922), edited by H. L. Mulder and B. F. B. Van de Velde-Schlick. Dordrecht: D. Reidel, 1979. Original edition "Kritizistische oder empiristische Deutung der neuen Physik?" *Kant-Studien* 26 (1921): 96–111.

Slowik, Edward. 2012. "On Structuralism's Multiple Paths through Spacetime Theories." *European Journal for Philosophy of Science* 2 (1): 45–66.

6   The Status of Newton's Laws

In explaining what he means by saying that Newton's laws are constitutive elements in science, Friedman describes the role they play in our understanding of the universal law of gravitation, which says that any two bodies in the universe will attract each other:

> Any two pieces of matter therefore experience accelerations towards one another in accordance with the very same law. But relative to what frame of reference are the accelerations in question defined? . . . The privileged frame of reference in which the law of gravitation is defined is what we now call an *inertial frame,* where an inertial frame of reference is simply one in which the Newtonian laws of motion hold (the center of mass frame of the solar system, for example, is a very close approximation to such a frame). It follows that without the Newtonian laws of mechanics the law of universal gravitation would not even make empirical sense, let alone give a correct account of the empirical phenomena. For the concept of universal acceleration that figures essentially in this law would then have no empirical meaning or application: we would simply have no idea what the relevant frame of reference might be in relation to which such accelerations are defined. (2001, 36)

We need Newton's laws in order to define an inertial frame, indeed they are taken to define an inertial frame. Newton's laws are thus preconditions in a very specific sense—they are a bridge between the mathematical structure of a theory and the empirical laws (Friedman 2001, 77). They are what Friedman calls, in the *Dynamics of Reason*, coordinating principles. He has now given up the idea of calling these coordinating principles as being too closely tied to an outdated theory of semantics, as the original problem was set out by Reichenbach and Schlick (Friedman 2010, 697–698).[1] Presumably, however, in this case, the Newtonian Laws do still define an inertial frame; it is just that they do not simply or directly connect the mathematics of the Newtonian theory to empirical experience.

There is a long-standing controversy concerning the epistemological status of Newton's laws of motion, that is, whether they are empirical, a priori, or conventional.[2] As one would expect of constitutive principles, their epistemological status is hard to pin down, at least in part because the laws can play different roles in different contexts. Even in the twentieth century, after Newton's laws were replaced with those found in the Special and General Theories of Relativity, a lively debate over the status of Newton's laws of motion took place in the philosophical literature, where the introduction of the first law was seen as an extremely important example of conceptual change (Hanson, 1958, 1965; Toulmin, 1961). The overthrow of Scholastic theories of motion by Newton was seen as instructive because shifts in what required explanation, such as continued motion, were implicated in the change in fundamental principles or laws. Understanding Newton's laws required a new way of thinking, not merely the acknowledgment of a new fact.

Newton's laws clearly play a foundational role in theories of motion, but the question of how they are justified is difficult. I argue that understanding these laws as constitutive is the most important fact about their epistemological status, not whether they are empirical or a priori. Although not a defense of a priori knowledge in the classical sense, the picture of science developed here is very different from that developed in Quinean holism in that categories of knowledge are strongly differentiated. The special role that Newton's laws play is not captured by saying simply that they are in the hard core of the web of belief but, rather, by the fact that Newton's laws made a science of motion possible. For example, one cannot formulate Newtonian mechanics without the concept of instantaneous velocity and one cannot define instantaneous velocity without the calculus.[3] Thus, the change from Scholastic to Newtonian theories of motion took place with new mathematics. By contrast, many scientific disputes, for example, the current dispute between two theories of human origins (the pathway out of Africa) can be settled when enough data are collected, without requiring any new mathematics or changes in concepts. Newton's principle of inertia changed the standard for what kinds of motion required explanation and what kinds were considered to be natural.

The idea of the constitutive elements in scientific theories provides a way of breaking away from the general debates over the existence of a priori knowledge (the debate between empiricists and rationalists) and of considering a much more interesting question that gets us much closer to an understanding of scientific practice, namely, how to describe the constitutive elements of a scientific theory. Some of the fundamental principles of a scientific theory are constitutive in the sense that they must be established prior to everything else and that they are very difficult to test or to eliminate, while the other parts of a physical theory can be tested and can be eliminated. The epistemological status of Newton's laws has attracted so much attention because they are an excellent example of principles (or laws) that are constitutive and are therefore very difficult to classify. If they are empirical, I claim, they are not simply so, a point that can be seen if we first consider the traditional debate between rationalism and empiricism.

RATIONALISM VERSUS EMPIRICISM

Since the distinctive trait of modern science is taken to be its combination of experiment and the application of reason, especially mathematics, to the study of nature, it may seem natural to give Newton's laws this role—a mixture of empirical and a priori. It seems impossible to justify classifying Newton as an empiricist, even if he did say, famously, that he will "frame no hypotheses." Indeed, many have argued that the most philosophically appealing aspect of Newton's methodology was his resistance to the extremes of empiricism and rationalism (Stein 1990), expressed in Newton's recognition of the need for both analysis and synthesis and for actually carrying out experiments (Hankins 1985). Still, the roles of reason and the senses in knowledge and in the formation of ideas is at stake in the philosophical debates over rationalism and empiricism precisely because of the inconsistent claims made about their roles in the then new science. These debates can be seen as having been born out of the methodological reflections on the relative roles of experiment and reason in creating the success of the new science; thus, it will be worthwhile to review briefly the debate of the role of reason and experiment here before considering specifically Newton's laws.[4]

The rationalist claims that some part of scientific knowledge about the physical world is a priori—known through reason or intellectual intuition—while the empiricist claims that knowledge about things in the world can only be obtained through experience. Both sides accept the ability of the mind to formulate and to understand representations of nature and acknowledge the role of perceptual knowledge in science and in everyday experience, however empiricists claim that reason is limited to what Hume calls the "relations of ideas," that is, the defining of one term by means of another or the discovery of the logical consequences of propositions (Hume 1966, IV). It is important to note that a rationalist need not be committed to a priori knowledge of the existence of anything, nor of the properties of any individual object, but rather only to general claims about the nature of things in the world. For example, a rationalist might claim that geometry expresses the real nature of space and the things in it, so any triangle (or even something that approximates a triangle) in nature must have certain characteristics that we can discover a priori. Once we know that a triangle is a three-sided figure, we can use pure reason to show that the sum of the three angles of any triangle must be equal to two right angles (Descartes 1984, 45; AT 64). Rational intuition is supposed to tell us how the world must be, since the general principles and laws that the rationalist claims to discover are not contingent facts.

The burden of proof for the rationalist is to explain what rational intuition is and why we should think that it will reliably tell us something about the world. The burden of proof for the empiricist in this debate, especially after Kant, is to show how science can exist without any a priori knowledge.

While Kant limited reason and acknowledged that experience is the main source of scientific knowledge, he also argued that there is a residual element of a priori synthetic knowledge, what we might now call the theoretical elements of a science that cannot be eliminated. For example, Kant argued that mathematics is both a priori and synthetic; that is, it tells us more than Hume's relations of ideas convey. Since mathematics is clearly central to science, it becomes a major stumbling block to the empiricist claim that there is no a priori element in science and suggests that a good way to investigate rationalism in science is to ask whether there are some elements of science that are constitutive. Mathematics, a few fundamental principles or laws of nature and other theoretical elements of science seem to be good candidates for a priori knowledge for which the empiricist will need to provide an account.

NEWTON'S FIRST LAW

In a standard formulation, Newton's first law says that a body free of impressed forces either remains at rest or else continues in uniform motion in a straight line. Pap claims that it is an empirical fact that Newton's first law is unrealizable because of the constant presence of friction and of universal gravitation (1946, 41–42). Friction slows bodies down and gravity diverts them from a straight path. Pap is, of course, aware of the possibility of reducing friction to a minimum and coming close, experimentally, to Newton's idealized law. Galileo accomplished such a reduction of friction with his inclined plane experiments. We can also observe some bodies traveling in deep space, far from the gravitational attraction of other bodies. I would note, however, that it appears that Pap's unrealizability argument can be strengthened by the a priori or conceptual point that the law cannot be empirically grounded if the body obeying the law and the observer are separate entities, given that they must affect each other. Thus, it seems impossible in principle to observe a body that is completely undisturbed by any other body. Many have argued that the existence of gravity and friction make Newton's first law at best hypothetical or subjunctive: If there were a body unimpeded by gravity or by friction it would continue at rest or in uniform rectilinear motion forever, but given Newton's theory of universal gravitation, it seems that no material body has ever been in such an unaffected state, let alone been observed in this state (Hanson 1965, 13).

Newton can explain what we observe with the first law in combination with gravity and/or with friction, but we never observe precisely what the law says that we should see. Therefore, Newton's first law is not a simple inductive generalization, given that it cannot be a simple inductive generalization if it is not in accord with observation. Newton's first law cannot easily be taken to be a priori, given that alternative theories do not seem to be ruled out a priori. Those who would claim that the laws have a priori grounding (in the classical sense—that they are synthetic a priori truths) will have a large burden of proof, I argue. The idea that the laws could all be definitions or analytic and thereby a priori will also flounder. Newton's first law is also strange, perhaps, in that for bodies at rest, the law accords with everyday observation, but for bodies in motion it does not. Bodies at rest remain at rest until they are disturbed by some external force, but bodies in motion appear to slow down and eventually stop, from an everyday (and a Scholastic) perspective. So, one implication of Newton's first law is that uniform motion is the same as rest, a claim that is consistent with what we now call Galilean relativity, that while traveling along with an object at constant velocity, one will notice no motion at all—the object can be considered to be at rest, relative to an inertial frame. How do with know what should be taken as natural motion and what needs to be explained? Perhaps it is easier to understand Newton's first law as a standard for the kind of motion that needs to be explained rather than as a law (Toulmin, 1961, 54–55; Ellis 1965). Newton claims that uniform rectilinear motion is the norm, with deviations from uniform rectilinear motion explained by external forces such as friction, gravity, or impact with another body, while the Scholastics hold that continued motion after impact needs to be explained. Thus, for example, the Scholastics contend that the motion of a cannonball can be explained not only by the initial impact of the explosion of gunpowder but also, according to the Scholastics, by the impetus that is imparted to the cannonball, which is necessary for its continued motion. I first consider some a priori arguments for the law of inertia, and then consider the argument that it is empirical.

A PRIORI ARGUMENTS

The Scholastic theory of impetus has often been ridiculed and called ad hoc, yet for someone making an a priori argument for Newton's first law, I maintain that discounting the Scholastic theory presents a real challenge. How does one know what kind of motion is natural? Why would motion in a straight line be considered natural rather than motion in the arc of a great circle, as Galileo is sometimes said to have argued?[5] The charge that the Scholastic theory of impetus is ad hoc can only be made when empirical considerations are used to test what happens to projectiles under conditions, for example, where friction varies. These arguments are not open to someone making a purely a priori case for Newton's first law. Furthermore, there are instances where the impetus theory seems to apply, such as in the case of the Atlatl, an ancient type of spear thrower, eventually replaced by the bow and arrow, which consists of a short stick with a hook used to launch a flexible spear or 'dart.' The mechanical advantage gained by the launcher was long understood, but this alone does not explain the tremendous power of this weapon, which was said to be the only thing that Cortez and his men feared while conquering the New World—it could pierce Spanish armor from a great distance. It turns out that the flexibility of the dart is the key. At launch, the dart bends, then straightens out and pushes against the stick, acting much like a spring, thus giving the dart much more momentum (Brown et al. 1996; Perkins 2013). Unlike the cannon, the Atlatl actually does impart a force to its projectile—a perfect example for the impetus theory, given that the launching of an Atlatl dart changes the dart itself. It does not remain passive like a cannon ball but, rather, plays a part in its own motion while it is being launched. While Newtonian theory can account, of course, for this case too, the Atlatl provides an example in which Newton's theory of inertia is not incorrect, but rather, it is not such an obviously superior theory to the Scholastic theory that came before it.

Russell Norwood Hanson rehearses Galileo's incline ramp experiments to reconstruct an interesting a priori argument for the law of inertia. Galileo found that a ball that was rolled down an inclined plane and then back up another inclined plane would roll up to (almost) the same height at which it began, a motion that is analogous to a pendulum. He then argued that on a completely frictionless surface, the height reached would be exactly the same height at which the ball started and that therefore the correct interpretation is that the ball will always reach its original height unless its motion is disturbed by an external force. As the angle of the second ramp is lowered, the ball must travel farther to reach its original height, and that as the angle of the ramp approaches zero degrees, the distance traveled will approach infinity. Therefore, we must assume that on a flat surface, a ball encountering no resistance will continue at the same speed indefinitely (Hanson 1965, 9–11). Thus, Galileo provides a thought experiment that extends his incline plane experiments to make an a priori argument for the law of inertia. However, we must remind ourselves that we do not know what happens on completely frictionless surfaces. More important, any effects of friction will be increased indefinitely as the distance that the ball must roll is extended. Any amount of friction will make it harder and harder for the ball to reach its original height if it has to travel a greater distance; therefore, the friction to be overcome is infinite if the distance to be traveled is infinite. Galileo's argument does show what happens if friction is reduced but not what happens when friction is zero.

Another a priori argument comes from Kant, of course, who defends the view that Newton's laws were synthetic and a priori, but in the end his position is a version of the constitutive theories I consider in this book. At least in the critical period, Kant understands the Newton's laws, or, more precisely, Kant's own formulations of them, as preconditions for any science of motion.[6] Michael Friedman expresses Kant's view as follows:

> Kant, since he rejects absolute space, conceives the laws of motion rather as conditions under which alone the concept of true motion has meaning: that is, the true motions are just those that satisfy the laws of

motion. . . . Kant thus views the laws of motion as definitive or constitutive of the spatio-temporal framework of Newtonian theory, and this, in the end, is why they count as a priori for him. (1992, 143)

Where Kant errs is first by seeing his laws of motion as apodictically certain, despite the fact that there are conceptual alternatives, as we saw above in the discussion of other a priori arguments for the first law. So while Kant is correct to see the laws as constitutive, his theory of the a priori (and apodictic) constitutive elements in science is not sustainable.

THE FIRST LAW AS EMPIRICAL

After Kant, Newton's first law frequently has been thought to be best interpreted as a definition and hence as an analytic truth. To be true, Newton's first law must be taken to refer to an inertial frame, given that bodies do not remain at rest or in uniform rectilinear motion in accelerated frames. However, we do not know in the beginning whether inertial frames exist or what they are. Newton solved this problem with absolute space, which he took to be the inertial frame against which all motion was measured. Barring acceptance of absolute space, many post-Kantian physicists and philosophers have suggested that the first law be used to define the concept of inertial frame. Like Kant, they understood the first law as specifying the meaning of the fundamental terms of Newtonian Mechanics, but unlike Kant, they understood the first law as an analytic statement. An inertial frame is precisely one in which particles obey Newton's first law. Earman and Friedman argue that this move is unnecessary, given that the concept of a straight line can be well defined in a four-dimensional version of Newtonian spacetime, since it is a four-dimensional Euclidean manifold. Thus, 'inertial frame' can be defined independently of the first law, so the law can be given empirical status. Earman and Friedman summarize their account of the status of Newton's laws as follows:

> What is the status of Newton's First Law? Again the answer is clear: on all of the above formulations it is an empirical law—it says something about the affine structure of space-time and how that structure constrains particle trajectories. (1973, 337)[7]

Essentially, Earman and Friedman interpret Newton's Laws within the context of the General Theory of Relativity, with the four-dimensional spacetime manifold as a mathematical description of physical spacetime. Since Earman and Friedman take a scientifically realist position about spacetime, Newton's Laws can be considered straightforwardly empirical. Even absolute space can be understood in a realist fashion that is as well defined and empirically grounded as physical spacetime in the General Theory of

Relativity, the current best theory of spacetime that we have. In viewing Newton's Laws as essentially empirical, Earman and Friedman represent the current thinking of many philosophers of science about spacetime. However, the remaining element of constitutive a priori in their account places their view closer to other theories that we have been considering here than they might have thought, given that the laws are constitutive.

NEWTON'S SECOND AND THIRD LAWS

As with the first law, Newton's second and third laws have also been taken not to be straightforward empirical claims by many philosophers and scientists. Poincaré gives an argument to show that Newton's second law must be considered to be a definition (1913, 97–99). The problem is that force and mass are defined in terms of each other, or perhaps more clearly, force and mass can only be measured as correlative terms, given by the law $F = ma$. Even if we assume that we know how to measure acceleration, we need to know how to measure mass and force independently of each other. Indeed, Poincaré says that there is a case to be made that the third law is also a definition, but he insists that there remains a sense in which the laws are making empirical claims:

> The principles of dynamics at first appeared to us as experimental truths, but we have been obliged to use them as definitions. It is *by definition* that force is equal to the product of mass by acceleration; here, then, is a principle which is henceforth beyond the reach of any further experiment. It is in the same way by definition that action is equal to reaction.
>
> But then, it will be said, these unverifiable principles are absolutely devoid of any significance; experiment cannot contradict them; but they can teach us nothing useful; then what is the use of studying dynamics?
>
> This over-hasty condemnation would be unjust. There is not in nature any system *perfectly* isolated, perfectly removed from all external action; but there are systems *almost* isolated.
>
> If such a system be observed, one may study not only the relative motion of its various parts one in reference to another, but also the motion of its center of gravity in reference to the other parts of the universe. We ascertain then that the motion of this center of gravity is *almost* rectilinear and uniform, in conformity with Newton's third law. (1913, 102–103, emphasis in original)

In Friedman's formulation quoted at the beginning of this chapter, it is the law of gravity that makes the empirical connection between the Newtonian laws of motion and what we can measure.

Hanson also discusses Newton's second law, showing that the second law is used in different ways in different contexts and can, he claims, be interpreted in very different ways because of these different uses. Hanson begins his discussion of Newton's second law of motion by replacing the 'a' in F = ma by the derivative of velocity, that is, of course, defining acceleration as the rate of change in velocity over time. He then replaces this again with second derivative, because velocity is the rate of change of position over time. So he obtains (with m assumed to be constant)

$$F = ma = m(dv/dt) = m(d^2s/dt^2)$$

and uses the second derivative in all further discussion (Hanson 1958, 99). In total Hanson considers nine separate accounts of the second law, if you count the main headings and subheadings:

'$F = m(d^2s/dt^2)$' has many distinct uses within mechanics. Consider these accounts:

1. F is *defined* as $m(d^2s/dt^2)$. In dynamics that is what 'F' means. It would be self-contradictory to treat 'F' as if it were not strictly replaceable by '$m(d^2s/dt^2)$'. (This is like our earlier examples.)
2. It is psychologically inconceivable that F should be other than $m(d^2s/dt^2)$. A world in which this did not obtain might as a matter of strict logic be possible, but it is not a world of which any consistent idea can be formed. On this equation rests all macrophysical knowledge. Were the world not truly described thus, the system, so useful in dealing with machines, tides, navigation and the heavens would crash into unthinkable chaos.
3. Perhaps, despite all appearances, $F = m(d^2s/dt^2)$ is false—unable adequately to describe physical events. Perhaps another set of conceptions could be substituted. Nonetheless this would be unsettling. $F = m(d^2s/dt^2)$ facilitates the collection and organization of a mountain of facts and theory. It patterns our ideas of physical events coherently and logically. So the second law, though empirical, cannot be falsifiable in any ordinary way, as are the statements which follow from initial conditions in accordance with this law.
4. $F = m(d^2s/dt^2)$ summarizes a large body of experience, observations, and experiments of mechanical phenomena. It is as liable to upset as any other factual statement. Disconfirmatory evidence may turn up tomorrow. Then we should simply write off $F = m(d^2s/dt^2)$ as false.
5. $F = m(d^2s/dt^2)$ is not a statement at all, hence not true, false, analytic, or synthetic. It asserts nothing. It is either:

   (a) a rule, or schema, by the use of which one can infer from initial conditions; or
   (b) a technique for measuring force, or acceleration, or mass; or
   (c) a principle of instrument construction—to use such an instrument is to accept $F = m(d^2s/dt^2)$, and no result of an experiment in which this instrument was used could falsify the law; or

(d) a convention, one of many ways of construing the phenomena of statics, dynamics, ballistics and astronomy; or

(e) 'F = m(d²s/dt²)' demarcates the notation we accept to deal with macrophysical mechanics. Our concern here, (a)–(e), is not with the truth or falsity of the second law. We are interested only in the utility of F = m(d²s/dt²) as a tool for controlling and thinking about dynamical phenomena.

The actual uses of 'F = m(d²s/dt²)' will support each of these accounts. This means not just that among physicists there have been spokesmen for each of these interpretations, but that a particular physicist on a single day in the laboratory may use the sentence 'F = m(d²s/dt²)' in all the ways above, from 1–5, without the slightest inconsistency. (1958, 99–100)

I do not discuss all of these in turn. However, if Hanson is right, he has provided good grounds for thinking that there is something special about the epistemological role being played by Newton's second law of motion. I want to highlight the idea presented in (5), that the law is not a statement at all and hence is neither true nor false and neither analytic nor synthetic. We saw the idea that some constitutive principles are neither true nor false in Poincaré, and we will see this idea taken up again by both Ian Hacking and Michael Friedman. The idea is that the constitutive principles are not claims in the ordinary sense, though they certainly appear to be. Instead, they allow a scientific practice to take place that does make claims.

BACK TO PAP'S FUNCTIONAL THEORY OF THE A PRIORI

In his discussion of the Newtonian Laws of motion, Hanson also develops a critique of Pap's functional theory of the a priori. Much like in the case of Michael Friedman, this may be surprising, since Hanson is very much a fellow traveler with Pap, in that he thinks that the epistemological status of the laws of motion is not at all straightforward. He tries to distinguish himself from Pap by saying that while Pap is making a historical point, he is not:

The laws of physics, particle physics especially, are used sometimes so that disconfirmatory evidence is a conceptual possibility, and sometimes, as above, so that it is not. This is not the historical point that physical laws begin life as empirical generalizations, but (through repeated confirmations, and good service in theory and calculation) they graduate to being 'functionally *a priori*'. Lenzen and Pap mark this well; Broad concedes it, but insists that the 'cash value' of law statements always rests in their relation to observation; Poincaré demurs, on the grounds that the laws of physics must keep in touch with experience. But the possible orderings of experience are limitless; we force upon the subject matter of physics the ordering we choose. (Hanson 1958, 97)

I find it puzzling that Hanson interprets Pap's view as essentially that of Quine, namely, that laws, which begin their lives as empirical, are gradually over time taken to be so well established that they cannot be changed. Hanson contrasts this view with what is expressed in the last sentence of the quote and I take to be his own view, namely, that we (somehow) set up a conceptual scheme that orders our experience and then we describe nature in that scheme. However, there is nothing in Pap's account that requires that the change of status from an empirical law to a hardened principle happen over time. Furthermore, it is clear in Pap that the process is reversible; that is, a sentence that functions as an a priori principle in one context could be falsified empirically in another context. When Pap says that sentences can have different roles in different contexts, I do not take him to mean that the change can only be historical. True enough, many of the examples he gives are historical in nature, which is not surprising given that our big examples of conceptual change are historical, but there is no reason to think that the changes must be historical and unidirectional, as Hanson seems to assume. The quote that I just gave continues as follows, continuing his discussion of Pap and Lenzen:

> These authors regard the shift in a law's logic (meaning, use) as primarily of genetic interest. They agree that at any one stage in the development of physics a law is treated in just one way, as empirical or as 'functionally *a priori*': in 1687 the law of inertia was apparently nothing but an empirical extrapolation; but in 1894 it functioned mostly in an *a priori* way. But this attitude is inadequate. It derives from the belief that a law sentence can at a given time have but one type of use. But the first law sentence can express as many things named 'The Law of Inertia' as there are different uses to which the sentence can be put. Now, as in 1894 and in 1687, law sentences are used sometimes to express contingent propositions, sometimes rules, recommendations, prescriptions, regulations, conventions, sometimes *a priori* propositions (where a falsifying instance is unthinkable or psychologically inconceivable), and sometimes formally analytic statements (whose denials are self-contradictory). Few have appreciated the variety of uses to which law sentences can be put at any one time, indeed even in one experimental report. Consequently, they have supposed that what physicists call 'The Law of Inertia' is a single discrete, isolable proposition. It is in fact a family of statements, definitions and rules, all expressible via different uses of the first law sentence. Philosophers have tendered single-valued answers to a question which differs little from 'What is *the* use of rope?' Once having decided their answers, they have to deprecate other obvious and, for their points of view, awkward uses of dynamical law sentences. (Hanson 1958, 97–98)

The main critique of Pap is that he assumes (Hanson supposes) that "a law sentence can at a given time have but one type of use." By contrast, Hanson urges us to think of statements of laws in science as having multiple contemporaneous uses and interpretations, as well as multiple historical uses and interpretations. Again, I just do not see that this criticism applies to Pap's functional a priori. It is true that Pap's statement in his preface seems to say no more than the fact that a former empirical statement can be, over time, taken to be functionally a priori:

> If, however, our point of view is dynamic or developmental, we shall find that what were experimental laws at one stage come to function, in virtue of extensive confirmation by experience, as analytical rules or "conventions," in Poincaré's language, at a later stage. (1946, vii)

However, on the very next page he makes his point general and defines the functional a priori as a contextual theory—one in which statements can play different roles in different contexts—and here he makes no mention of the genetic or historical component:

> The theory of the *a priori* which will, in this essay, be presented and applied to physical principles, may be called *functional* in so far as the *a priori* is characterized in terms of functions which propositions may perform in existential inquiry, no matter whether they be, on formal grounds, classified as analytic or as synthetic. It may also be called *contextual;* for statements of the form "x is a priori" or "x is a posteriori" (where the admissible values of x are propositions) will be treated as elliptical or incomplete. A proposition which is a priori in one context of inquiry, may be a posteriori in another context. (Pap 1946, viii, emphasis in the original)

Thus, we do not have to attribute the more specifically historical theory to Pap as Hanson does. Indeed, when he specifically discusses Poincare's theory of the hardening of principles, Pap again puts the point generally in terms of context, not in terms of genetic development (1946, 24).

In general I claim that there is more to Pap's theory of the functional a priori than his critics have allowed. To some extent what we see in these critiques is a kind of positioning, explaining how the critics view is different from and better than Pap's. I suppose that is to be expected, though it is ironic that in the case of Pap, both Friedman and Hanson reduce Pap's functional a priori to Quinean holism; that is, they would decline to count Pap's view as a theory of the constitutive elements in science at all and push it over to the opposing point of view. One might imagine that they would want to see themselves and Pap as allies.

THE ROLE OF MATHEMATICS

We have seen that a key issue in all of the discussion above is the role of mathematics in Newtonian theory, especially in the interpretation of Newton's Laws of Motion. The mathematical entities that the laws invoke seem not to be empirical, no more so than any other pure mathematical entity given that we have no causal relation to them. They are, in fact, the constitutive elements, in Kant's sense, of Newton's theory. Pap notes that in contrast to the Scholastic theory of impetus, Galileo explained the motion of bodies geometrically, dividing the forces acting on a projectile into two components—friction acting in the direct opposite direction as the straight inertial path of the projectile and gravity acting orthogonally to pull the projectile to the earth. These two forces can then be combined as the cross product of two vectors:

> Galileo saw a simpler way of explaining the parabolic trajectory of projectiles. The law of inertia, originally established by extrapolation from experiment, thus functions as a rule for the geometric construction of actual motions. It is a statement about a hypothetical component of actual motions, just as the law of the parallelogram of vectors assumes the causal efficacy of vector components to which not isolated existence can be ascribed. . . . no given component can be said to *exist* physically, unless it can be identified with an approximately isolable physical force of gravity. . . . In Kantian language, it is synthetic a priori in the sense of being a "constitutive condition" of mechanics: motion is a possible object of mechanics only in so far as it is geometrically constructible . . . (Pap 1946, 43)

Earman and Friedman admit similar kinds of geometrical objects play a fundamental role in Newtonian Mechanics; they actually replace the physical, causal process with a mathematical one. Since the bodies in question become mathematical, they cannot be (noncontroversially) empirically justified:

> Thus, in using Kepler's Laws to find the force of attraction exerted by the sun, Newton replaces the continuously acting force by a sequence of impulses acting at intervals Δt, replaces the actual path of the planet by a sequence of inertial paths, and passes to the limits as Δt goes to zero. (Earman and Friedman 1973, 331)

Classical Newtonian mechanics also uses a single mathematical point as the center of mass of a body. All of these replacements could be seen as mere idealizations, but they are, nevertheless, examples of an important point: Newtonian Mechanics would be impossible even to formulate without these mathematical tools. It follows that the mathematics used in Newtonian Mechanics is ontologically and epistemological prior to the physical claims made in the theory and therefore functionally a priori in the sense that it is a precondition, or in Kantian language, it is constitutive.

THE CONSTITUTIVE ELEMENTS IN SCIENCE

Newton's laws were seen as good examples of a pragmatic or changeable a priori, a view that was most fully explored by Pap (1946) in his dissertation.

The key element of this view is similar to Reichenbach in retaining Kant's idea of the constitutive aspect of a priori knowledge while removing the idea of a priori knowledge as necessarily true (Reichenbach [1920] 1965, 48). Thus, fundamental conceptual change in science is seen as changes in fundamental principles that are necessary preconditions to scientific knowledge—principles that had been seen as synthetic and a priori by Kant—but eliminating their necessity, given that they have in fact changed during a scientific revolution. These elements of physical theory have a unique epistemological status as a functionally a priori part of our physical theory, given that they are preconditions. Pap takes Poincaré's conventionalism for granted as obviously correct, which it is not. He also assumes as well that mathematics is conventional, apparently taking a formalist position in the philosophy of mathematics. In fact, showing that there are mathematical entities that replace physical ones just pushes the question of the status of Newton's Laws back to the question of the status of mathematics, something I consider in Chapter 8.

Background theory and context determine what is a priori and what is empirical, that is, which parts of the theory function as a priori knowledge, however, the background theory itself is determined empirically, in a broad sense of the term, by which I mean that while there is no single test that shows that the General Theory of Relativity is superior to Newtonian theories of motion and gravity and that the choice between the two theories is in part a result of the empirical consequences of each. Thus, the Newtonian theories of space and of motion are empirical theories, while the metric of space is Euclidean in constitutive and functions as a priori knowledge within Newtonian theory, which is why I argue that Newton's Laws are still empirical, albeit only in a very indirect and abstract sense. There are (at least) two stages of the justification of a theory, a constitutive stage and an empirical stage (Creath 2010). The fact that the empirical justification of a theory is only indirect does not make it less empirical, just empirical in a different sense.

The constitutive elements of a physical theory can sometimes be determined empirically, which means that a theory can be grounded empirically but still contain constitutive elements that cannot be directly tested and without which the theory could not be stated. On one hand, we could start from the constitutive functionally a priori elements, pointing out that these are necessary, while on the other hand, we could start from the whole empirical theory, pointing out that different empirical theories have different constitutive a priori elements. The constitutive elements are therefore not determined a priori, nor conventionally, but rather empirically as embedded elements in a physical theory. For example, Newton's laws are necessary preconditions of Newton's theory of motion, seeming to make them a priori, but at the same time, they are superseded by Einstein's theory of Relativity. Einstein's fundamental laws and the applications of mathematics that he uses are not only justified conceptually but also, in part, by the empirical success of the General Theory of Relativity.

In his more recent work, Friedman (2001) adopts a viewpoint of the relativized a priori, which should be inconsistent with his early position staked out with Earman, away from which Friedman certainly sounds like he has moved. I argue that he goes too far, however, in attempting to defend a special position for philosophy with the notion of a dynamic a priori, given that the constitutive elements are fully part of science. It would be odd indeed to say that anything that science requires is nevertheless separate and not part of science.

NOTES

1. See Oberdan (2009, 197–198) on some of the problems related to this issue.
2. I use the term *law* in order to maintain the traditional terminology without taking any particular stand here on the nature of laws. I do not attempt to cite all of the papers or to rehearse all of the arguments on Newton's Laws but, rather, focus on a particular treatment of constitutive elements in science. See Whitrow (1950), Pap (1946), Ellis (1965), Shapere (1967), and Earman and Friedman (1973) for more complete surveys.
3. Earman and Friedman deny that this is the case for Newton's original formulation of the first law (1973, 331). I intend *Newtonian* to refer to the standard development of Newtonian mechanics, not Newton's own formulations.
4. I expand considerably on the material in this section in Stump (2005).
5. Galileo was interpreted this way by Koyré and his interpretation was rejected by Drake. For an excellent recent analysis of the dispute, see Maarten Van Dyck (2010).
6. See Watkins (1998) for a discussion of Kant's formulations of Newton's laws and for the arguments that Kant gives in the precritical period.
7. Earman and Friedman continue this quotation with the comment that Newton's first law "is superfluous since it is a consequence of the Second Law" (1973, 337). Pap makes the same point (1946, 48) and indeed it is true that in the modern formulation the first law is superfluous. However, Earman and Friedman go on to argue in section 4 of their paper that in fact the First Law is not redundant, given the role that it has to play in Newton's theory (1973, 345–346).

BIBLIOGRAPHY

Brown, M. M., D. Pritchard, P. Chauvaux, and C. A. Hoyt. (2013). *Atlatl Literature Resources*. Available from http://www.flight-toys.com/atlatl/atlatl_books.html.
Creath, Richard. 2010. "The Construction of Reason: Kant, Carnap, Kuhn, and Beyond." In *Discourse on a New Method: Reinvigorating the Marriage of History and Philosophy of Science*, edited by M. Domski and M. Dickson. Chicago and La Salle, IL: Open Court, 493–509.
Descartes, Rene. 1984. "Meditations on First Philosophy." In, *The Philosophical Writings of Descartes*, vol. II, translated by J. Cottingham, R. Stoothoff, and D. Murdoch. Cambridge and New York: Cambridge University Press, 1–62.
Earman, John, and Michael Friedman. 1973. "The Meaning and Status of Newton's Law of Inertia and the Nature of Gravitational Forces." *Philosophy of Science* 40: 329–359.

Ellis, B. D. 1965. "The Origin and Nature of Newton's Laws of Motion." In *Beyond the Edge of Certainty: Essays in Contemporary Science and Philosophy*, edited by R. G. Colodny. Englewood Cliffs, NJ: Prentice Hall, 29–68.

Friedman, Michael. 1992. *Kant and the Exact Sciences*. Cambridge, MA: Harvard University Press.

———. 2001. *Dynamics of Reason: The 1999 Kant Lectures at Stanford University*. Stanford, CA: CSLI Publications.

———. 2010. "Synthetic History Reconsidered." In *Discourse on a New Method: Reinvigorating the Marriage of History and Philosophy of Science*, edited by M. Domski and M. Dickson. Chicago and La Salle, IL: Open Court, 571–813.

Hankins, T. L. 1985. *Science and the Enlightenment*. Cambridge: Cambridge University Press.

Hanson, Norwood Russell. 1958. *Patterns of Discovery: An Inquiry into the Conceptual Foundations of Science*. Cambridge: Cambridge University Press.

———. 1965. "Newton's First Law: A Philosopher's Door into Natural Philosophy." In *Beyond the Edge of Certainty: Essays in Contemporary Science and Philosophy*, edited by R. G. Colodny. Englewood Cliffs, NJ: Prentice Hall, 6–28.

Hume, David. 1966. *Enquiries Concerning the Human Understanding and Concerning the Principles of Morals*. Oxford: Clarendon Press.

Oberdan, Thomas. 2009. "Geometry, Convention, and the Relativized Apriori: The Schlick–Reichenbach Correspondence." In *Stationen. Dem Philosophen und Physiker Moritz Schlick zum 125. Geburtstag*, edited by F. Stadler, H. J. Wendel, and E. Glassner. Vienna: Springer, 186–211.

Pap, Arthur. 1946. *The a priori in physical theory*. New York: King's Crown Press.

Perkins, B. 2013. *Atlatl & Dart Mechanics*. Available from http://www.atlatl.com/mechanics.php.

Poincaré, Henri. 1913. *The Foundations of Science: Science and Hypothesis, The Value of Science, Science and Method*. New York: The Science Press.

Reichenbach, Hans. (1920) 1965. *The Theory of Relativity and A Priori Knowledge*. Berkeley: University of California Press.

Shapere, Dudley. 1967. "Newtonian Mechanics and Mechanical Explanation." In *Encyclopedia of Philosophy*, edited by P. Edwards. New York: Macmillan, 115–119.

Stein, Howard. 1990. "On Locke, 'the Great Huygens, and the Incomparable Mr. Newton.'" In *Philosophical Perspectives on Newtonian Science*, edited by R. I. G. Hughes. Cambridge, MA: MIT Press, 17–47.

Stump, David J. 2005. "Rationalism in Science." In *Blackwell Companion to Rationalism*, edited by A. Nelson. Oxford: Blackwell, 408–424.

Toulmin, Stephen. 1961. *Foresight and Understanding: An Enquiry into the Aims of Science*. Bloomington: Indiana University Press.

Van Dyck, Maarten. 2010. "Haunted by Circular Motions: Koyré and Drake on Metaphysics and the Scientific Revolution." Presented at the HOPOS conference, Budapest, Hungary.

Watkins, Eric. 1998. "Kant's Justification of the Laws of Mechanics." *Studies in History and Philosophy of Science* 29 (4), 539–560.

Whitrow, G. J. 1950. "On the Foundations of Dynamics." *The British Journal for the Philosophy of Science* I: 92–107.

7 Conceptual Change and the Constitutive Elements of Science
Kuhn, Friedman, and Hacking

KUHN'S PARADIGMS

In an interview near the end of his life, Thomas Kuhn said, "I go around explaining my own position saying that I am a Kantian with moveable categories" (2000, 264). Indeed, Kuhn's image of science emphasizes the constitutive elements in science, and this is hardly surprising, given that Kuhn is trying to understand conceptual change in science as the basis for scientific revolutions. Earlier, Kuhn gave a little more detail of what he meant:

> Though it is a more articulated source of constitutive categories, my structured lexicon resembles Kant's a priori when the latter is taken in its [Reichenbach's] second, relativized sense. Both are constitutive of *possible experience* of the world, but neither dictates what that experience must be. (1993, 331)

He makes the same basic point at the end of his Philosophy of Science Association Presidential Address, that there is a framework and empirical content within that framework (Kuhn 2000, 104). Let us briefly review Kuhn's main points in *The Structure of Scientific Revolutions* (1962) in order to set the context for his remarks about the Kantian a priori.

Kuhn claims that history can provide us with a radically different image of science from that typically found in science textbooks and in philosophy of science. History, however, must look at older periods of science in their own terms, rather than from our modern perspective. Kuhn justifies this outlook by claiming that it is best historical practice, that is, it is the best way to write history and is the outlook that professional historians adopt. What Kuhn found by looking at the history of science was stable periods of what he called normal science, interrupted by revolutions. The revolutions were the focus of *Structure*, given that Kuhn was interested in the conceptual change that they embody. Kuhn found many revolutionary changes in science and found that these constituted significant breaks from past scientific practice.

Paradigm is the key term in Kuhn's very influential book. As is frequently the case when new ideas are presented, Kuhn took an existing term and gave it a specialized meaning. The term *paradigm* now occurs frequently in every kind of discourse, usually to mean something like 'way of thinking' or 'approach to a problem.' Kuhn has generally been given credit for introducing this usage, but the way that *paradigm* is popularly used misses a central aspect of his argument. Kuhn emphasizes that a paradigm cannot be reduced to a set of beliefs or to a list of rules and indeed, that a paradigm cannot be put completely into words. Scientists have to learn by doing, both by thinking in terms of the concepts that are used in a particular science and by physically manipulating material to create phenomena. Kuhn later separated the paradigm (as a model) from the "disciplinary matrix" that guides the activity of normal science. Later still, he spoke of "structured lexicons" basically replacing the disciplinary matrix, but none of these changes makes a difference for the argument that is being presented here, so I describe Kuhn's views in terms of paradigms.

Paradigms organize scientific activity and makes normal science possible. Once a paradigm has been established, scientists can use it to do more specialized and esoteric work. Scientific work can be specialized because the general issues have been settled in the work that established the paradigm. Scientific work is esoteric because only those who have been trained in the research community that accepts the paradigm are able to understand it. Normal science is defined by a paradigm as a set of open problems and puzzles that are well defined enough to be recognized and worked on by a community of specialists. Methods, metaphysical commitments, theories, and aims are more or less shared by a research community, but there can be disagreement and ambiguity as well. By determination of facts (checking them with theories) and articulation of the paradigm (saying what it means and extending it by applying it to new areas of inquiry), normal science is cumulative and progressive.

Kuhn compares normal science to puzzle solving, emphasizing that the solutions found in normal science are expected and that the problems to be solved, the methods to be used, and so on are agreed on by scientists in a period of normal science. He also claims that normal science cannot be defined by a set of strict rules, since there is something left incomplete or implicit in the definition of a scientific practice. He insists that we can learn more about science by looking for the paradigms that scientists follow than by looking for a set of rules that they follow. Kuhn compresses his discussion of the centrality of the notion of paradigm into a single chapter titled "The Priority of Paradigms" (1962, 43–51). Paradigms have priority because there is nothing more basic by which 'paradigm' could be defined. In logical terminology, the word *paradigm* functions as a primitive term. Properties of paradigms can be given and examples of paradigms can be enumerated, but the word cannot be defined, no more than 'number' can be defined in arithmetic. Kuhn justifies his introduction of the term *paradigm* by arguing that, for the historian, it is a better organizational concept than any other. By looking for paradigms and changes in paradigms, the historian can classify scientists and historical periods in ways that lead to productive research and a better understanding of the history of science. Turning to philosophers to justify the indefinability of paradigms, Kuhn invokes Michael Polyani's idea of tacit knowledge and Wittgenstein's idea that some human activities cannot be captured by a set of rules (1962, 42), arguing that while paradigms cannot be reduced to a set of methods and beliefs, they are recognizable to the historian as the organizing principle underlying a period of normal science.

The root definition of *paradigm* as both pattern and example exhibits both sides of a classical philosophical debate over the nature of definition. Plato argued forcefully that providing examples is not adequate; the real definition of a term must specify what the examples have in common and thus explain why they all properly fall under the concept being defined. In the terminology of later philosophy, Plato argues that a definition must tell us the essence of a thing. For Plato, this is the *eidos*, the eternal form or idea to which all objects falling under a concept must conform. Kuhn sides with Hume and Wittgenstein in rejecting Plato's requirement that the essence be given in the proper definition of a concept. Like Hume, Kuhn argues that it is enough to say that the objects falling under a concept resemble one another in various aspects that can be specified and to take that resemblance as a starting point. Kuhn claims that paradigms do not have an essence, since there is always some disagreement and some difference in emphasis among scientists who are working under the same paradigm. In Wittgenstein's terminology, the historian can find a family resemblance among the views of these scientists, rather than single common set of beliefs and methods.

Kuhn also defends his view that paradigms cannot be reduced to a set of beliefs and rules of method by pointing out that scientists learn by working through concrete examples of problems, not by learning rules (1962, 46). Thus, learning a paradigm is more like learning a skill than like learning a body of knowledge, a point that Joseph Rouse (2003) has rightly emphasized. Kuhn is very close to using *paradigm* in its original meaning here, since the problems and solutions through which students learn are to be taken as patterns of scientific thought and work.

Anomalies are phenomena that a paradigm cannot solve. Every paradigm has them—to say otherwise would be to say that science could be complete. Sciences working under a particular paradigm may or may not be aware of anomalies and they may or may not take the anomalies seriously. However, when anomalies are noticed and begin to be taken seriously, the paradigm may be led to crisis. In particular, the scientific community may split at this point, with some of its members being inclined to search for new paradigms (or at least to be open to them) while others are content to continue working with the established paradigm. Kuhn presented Lavoisier and Priestly's dispute over oxygen and phlogiston as an example of this phenomenon.

Kuhn describes competing paradigms as incommensurable. He tends to describe this in terms of language, which is that there is no way to translate from one paradigm to another, so they are not directly comparable. As I have mentioned earlier, I am wary of arguments having to do with meaning. I think far better examples that make Kuhn's point have to do with standards of explanation or with a change in questions that are being asked. As I mentioned before, when Newton (and Galileo) propose a law of inertia, they profoundly change the question being asked, so the dispute over what needs to be explained cannot be settled by a simple empirical test. Newton's theory of motion is superior to Aristotelian impetus theory, but the two cannot be directly compared.

KUHN AND THE RATIONALITY OF SCIENCE

Kuhn argued that his use of the term *revolution* to describe changes in science is appropriate because, like political revolutions, scientific revolutions overturn existing rules and institutions in order to establish new ones. By definition, there can be no legal way to have a political revolution, since any changes that follow the processes of the old regime would merely be reform, not revolution. For Kuhn, the key point of the analogy between political and scientific revolutions is that in both cases, there are no rules than could help adjudicate between the two systems (1962, 93). The supporters of the old and the new paradigms will each follow their own methods, emphasize their own aims, and accept their own solutions to problems, without necessarily accepting any of the methods, aims, or standards of supporters of the other paradigm. In an influential paper that helped redirect criticism of Kuhn's book, Jerry Doppelt (1978) emphasized the apparent relativism of Kuhn's view, given that there is no right or wrong answer to the question of when an old paradigm should be abandoned and a new paradigm adopted. Lavoisier and Priestly independently discovered oxygen, but while Lavoisier used this discovery as a basis for a new chemistry, Priestly never accepted Lavoisier's revolution and maintained the old phlogiston paradigm instead. Kuhn argued that both of these famous scientists were acting reasonably (1962, 93). Nothing can force a scientist to change paradigms, according to Kuhn, because a scientist can always find a way either to incorporate new data into the existing paradigm or to show why the new data can be dismissed as unimportant from the point of view of the existing paradigm. It is important to note, however, that Kuhn is not saying that anyone can believe anything they wish. Paradigms must be well developed and cover a wide range of phenomena. It is not easy to develop a new science that will justify the overthrow of an established paradigm.

Note that there are two kinds of issues raised by Kuhn—ontological and epistemological. Kuhn's claims about the world changes and about the incommensurability of paradigms are more ontological in nature, while his rejection of the usual tests of scientific theories and replacing them with more subjective or social explanations of theory choice are more epistemological in nature. Rather than promoting relativism, Kuhn saw himself as rejecting particular philosophical accounts of science and he accepts the blame for overblown rhetoric that misled his readers to think that he had set out to undermine the cognitive authority of science (2000, 228). He criticizes the idea of confirming scientific theories and comparing how well they are confirmed, a view of science associated with Carnap, and he criticizes the idea of testing scientific theories to show that one theory is false, a view of science associated with Popper and Hempel (Kuhn 1962, 145–147). Both of these views require that a body of neutral evidence be available to scientists, a position that Kuhn disputes because, he claims, all evidence is acquired on the basis of a paradigm and is therefore an element of that paradigm. In chapter 12 of *Structure*, Kuhn says that he will explain "[w]hat causes the group [of professional scientists] to abandon one tradition of normal research in favor of another?" (1962, 144). Since he has already said that evidence and arguments can only be given within a paradigm and that there is no such thing as neutral evidence or universal methods, it appears that the best arguments that can be given are, in some sense, circular or question begging. In other words, the evidence presented may only be accepted by those who have already adopted the paradigm in question—a view that sounds like relativism.

In the postscript, when Kuhn defends himself from the charge that he is a relativist, he argues that the kind of neutral evidence that would give one the ability to determine the truth of a paradigm is simply impossible to find:

> There is, I think, no theory-independent way to reconstruct phrases like 'really there'; the notion of a match between the ontology of a theory and its "real" counterpart in nature now seems to me illusive in principle. . . . I do not doubt, for example, the Newton's mechanics improves on Aristotle's and that Einstein's improves on Newton's as instruments for puzzle solving. But I can see in the succession no coherent direction of ontological development. (1962, 206, 173)

Therefore, Kuhn can claim to have explained as much as one can explain when it comes to choice of paradigms. Later paradigms do not get us closer to the truth, even if they seem obviously superior to previous paradigms. His position cannot be considered relativist, he argues, because there is no alternative way of thinking about paradigm choice.

What does Kuhn mean by saying that he is a Kantian with moveable categories? Kuhn says that his paradigms (or disciplinary matrixes or structured lexicons) are conditions of the possibility (in Kantian language) of establishing a science. A paradigm must be in place for normal science to exist, just as there is a hierarchy of sciences in Poincaré, each presupposing what comes before, the elements of a paradigm—methods, aims, ontology, lexicon, and so on—must be established so that scientists will know what to do. These elements are functionally a priori in the sense of being necessary preconditions. Kuhn recognized that his view was thus similar to Reichenbach's early view, in that the elements of a paradigm that had been fixed and universal for Kant are precisely what changes during a scientific revolution.

FRIEDMAN'S RESCUE OF THE RATIONALITY OF SCIENCE

Starting with papers on Reichenbach's early view, in the *Dynamics of Reason* and in following papers, Michael Friedman has done more than anyone to bring the idea of relativized constitutive elements in science to light and to defend it (Friedman, 2001, 2002, 2003, 2004, 2005a, 2005b, 2006, 2008a, 2008b, 2009, 2010a, 2010b, 2011, 2012). Indeed, it appears that it is Friedman who made Kuhn aware of this aspect of his own work (2000, 245). After setting out Friedman's evolving version of the dynamic a priori, I argue that there is no need to go as far as Friedman does in order to show the continuity and objectivity of science. Indeed, I argue that a there are already views that meet the challenge that Friedman sees in Kuhn's work, that is, the challenge of how to avoid the charge that we are left with less than fully rational reasons for changes of scientific theories during a revolution and left therefore with a kind of relativism.

As in all the theories of the constitutive elements in science, Friedman distinguishes between empirical laws and constitutive (a priori) principles. The example that he has worked through in great detail is the history of mechanics and gravitational theory from Newton to Einstein. As Friedman shows, Newton's laws (or their relativistic counterparts) and the mathematics used in physics are the constitutive elements, while Newton's law of gravity and Einstein's field equations are empirical. The constitutive elements, of course, changed from Newtonian theory to Einstein's—space is no longer Euclidean but, rather, has variable curvature and Newton's second law receives a relativistic "correction" so that it is consistent with the principle that nothing can travel faster than the speed of light:

> Thus, for example, whereas Euclidean geometry and the Newtonian laws of motion were indeed necessary presuppositions for the empirical meaning and application of the Newtonian theory of universal gravitation (and they were therefore constitutively a priori in this context), the radically new mathematical and physical framework consisting of the Riemannian theory of manifolds and the principle of equivalence defines an analogous system of necessary presuppositions in general relativity. (Friedman 2008b, 251)

Thus, some constitutive principles are necessarily in place, but they can change rather dramatically from one scientific theory to another. Friedman calls these constitutive elements the dynamic a priori. His theory of the constitutive elements in science, like other such theories that we have studied, is designed to account for scientific revolutions and the conceptual change that they entail.

What is unique about Friedman's account is the special role that he gives to philosophy. He sees the constitutive and the empirical elements in scientific theories as layers and adds a third—philosophical meta-paradigms or meta-frameworks. These are the intellectual context in which scientists work and they are essential to the development of the new constitutive or a priori elements of a theory:

> Moreover, what makes the latter framework constitutively a priori in this new context is precisely the circumstance that Einstein was only able to arrive at it in the first place by self-consciously situating himself within the earlier tradition of scientific philosophy represented (especially) by Helmholtz and Poincaré—just as this tradition, in turn, had earlier self-consciously situated itself against the background of the original version of transcendental philosophy first articulated by Kant. (Friedman 2008b, 251)

I have no doubt that philosophy plays big role in certain kinds of scientific disputes, but I want to resist the idea that philosophy plays the role that Friedman describes. In the first place, science is already continuous enough. In the case of relativity, there are transitional figures who are neither purely Newtonian nor purely Einsteinian—I have figures such as Poincaré and Lorentz in mind. In the second place, I want to resist at all costs the idea that philosophy can be set up as an independent and prior discipline that somehow grounds the sciences. On this point, I am fully in agreement with the naturalists—there is no first philosophy. Friedman is on record opposing naturalism, though what he seems to have it mind is mainly the Quinean version. There are many forms of naturalism, but what I like about some versions of the naturalist position is that science and philosophy should be seen as continuous, that is, as mutually influencing each other. Friedman should be sympathetic with that, since much of his work has been precisely on the interaction between philosophy and science, so I find it rather surprising that he gives philosophy a special role in the *Dynamics of Reason*.[1]

What exactly is the problem that this philosophical layer is supposed to solve? The answer is that it is supposed to provide a continuity that is lacking in the sciences. Friedman sets out this problem as stemming directly from Kuhn's account of scientific revolutions:

> [S]ince there appear to be no generally agreed upon logical rules governing the transition to a revolutionary new scientific paradigm or conceptual framework, there would seem to be no sense left in which such

a transition can still be viewed as rational, as based on good reasons. Non-rational factors, having more to do with persuasion or conversion than rational argument, must necessarily be called in to explain the transition in question. And, on the other hand, since only such a non-rational commitment of the scientific community as a whole can explain the acceptance of a particular scientific paradigm at a given time, it would appear that the only notion of scientific rationality we have left is a relativized, sociological one according to which all there ultimately is to scientific rationality—and thus to scientific knowledge—is the otherwise arbitrary commitment of some particular social community or group to one particular paradigm or framework rather than another. (2001, 47–48)

Thus, Friedman thinks that there is a problem left over from Kuhn's account of science—the reasons given to justify a scientific claim are always relative to a paradigm so we must turn to other kinds of explanations for why scientists adopt the views that they do, which leaves the door open for relativism and social construction. Friedman proposes a solution to this problem: first, that we see more continuity in the shifts than Kuhn might allow and, second, that philosophy forms a background against which the scientific advances seem rational and continuous. I agree with the first point, but not the second:

How, more specifically, can the proposal of a radically new conceptual framework be, nonetheless, both rational and responsible? In accordance with our threefold perspective on interparadigm convergence we can now say the following: first, that the new conceptual framework or paradigm should contain the previous constitutive framework as an approximate limiting case, holding in precisely defined special conditions; second, that the new constitutive principles should also evolve continuously out of the old constitutive principles, by a series of natural transformations; and third, that this process of continuing conceptual transformation should be motivated and sustained by an appropriate new philosophical meta-framework, which, in particular, interacts productively with both older philosophical meta-frameworks and new developments taking place in the sciences themselves. (2001, 66)

In the first place, there is a real question whether there is any such philosophical meta-framework, especially one that meets Friedman's goal of obtaining "an ideal state of maximally comprehensive communicative rationality in which all participants in the ideal community of inquiry agree on a common set of truly universal, trans-historical constitutive principles" (2001, 67). As Hasok Chang (2008) points out, there is a question of whether this should even be a goal or whether some kind of pluralism may be preferable, but my point is that there are no truly universal, transhistorical principles, especially not in philosophy. There are no such things, and furthermore, we do not need them to be rational.[2] We can make rational choices with rather mundane principles that are located in a particular time and place. There is no guarantee, of course, that we will not sometimes be led to an impasse or a situation where we do not have enough evidence to settle a dispute or to answer a question. In such cases we simply have to wait for more evidence to accumulate over time.

Put most strongly, I would argue that there is simply no problem to be addressed by recourse to a universal realm of philosophy—scientific revolutions can be rational even without the continuity that this layer of discourse provides. There is no reason to accept Kuhn's view on face value, as Friedman apparently has. A nonfoundationalist account of science has the resources to rationally ground theory choice in scientific revolutions. I have in mind views such as those found in the works of Larry Laudan (1977, 1987) and Dudley Shapere (1980, 1986), and I defend them in some detail against the charge of relativism in an early article (Stump 1991). I briefly summarize the argument here.

Three issues seem to be central to criticisms of nonfoundational accounts of science. First, the lack of absolutes—the fact that the methods and aims, as well as the content of science have changed—is taken to lead to relativism. I claim that a proper understanding of fallibilism and of the resources of the nonfoundational model disposes of this objection. Each of our judgments can be grounded without reliance on any fixed standard. Kuhn is wrong to imply that reasons given according to a paradigm are not good enough reasons. Several philosophers of science have argued that there can be independent constraints on theory choice even if we accept the view that all observation is theory-laden or built into a paradigm (Hacking 1983, 183; Shapere 1984; Galison 1987, 1988, 1989; Kosso 1988, 1989). These constraints result in objective choice because the theories that are presupposed can be epistemically independent of those under test. Thus, reliance on independent constraints opens a path to a nonfoundational account of objective knowledge and again underscores the idea that while scientists set up a practice that is situated at a particular time and place, is communicated with a particular vocabulary, and includes presuppositions, they cannot control what happens when they interact with nature. Of course, the scientist have to then interpret what happened when they interacted with nature, but there are limits on how far interpretation can be carried.

Second, the difficulty of rationally justifying cognitive aims is thought to lead to relativism. The traditional model of science is that facts, methods and aims form a hierarchy. Methodology determines what counts as a fact and the basic aims of science determine (through philosophical analysis) what counts as proper methodology. This model leads to a kind of relativism in the views of Popper and some of the Logical Positivists since they see aims and methods as conventional. Acceptance of the traditional hierarchy leads to a false dilemma. If we accept the traditional hierarchy, the only avenue left for grounding methodological principles and aims is a priori philosophical analysis. Thus, if the traditionalists are right, we must either hold onto the discredited a priorism or else embrace relativism. Nonfoundationalists have turned to naturalism to justify cognitive aims and the methods of science. Epistemological naturalists reject the hierarchy and attempt to avoid conventionalism by claiming that scientific facts can influence the choice of methods and aims. There is no strict hierarchy but, rather, an interaction among theories, methods, and aims. While naturalism exists in several forms, a general formulation will be adequate for the discussion here. Traditionalists hold that methodology is an autonomous, a priori discipline, while naturalists hold that methods and cognitive aims can be informed by scientific knowledge. There are variations on how autonomous or how connected to science methodology will be and on whether naturalism can support normative claims, but I do not think that these disputes are important in rejecting foundationalism.

Finally, the nonfoundational account must block the Cartesian demand that all knowledge claims be assessed at the same time. The Cartesian demand undercuts all of the fallible presuppositions on which nonfoundationalists would like to ground their judgments, leaving an irrationalism, which says that there are no ultimate grounds for our judgments and a demand for a return to foundationalism and the impossible task of finding some absolute foundation. I claim that the charge of relativism can be blocked without degradation of epistemic standards. Holism makes it seem as if theories, methods, and aims are tightly connected in a single matrix such that no theories or aims are independent and makes the charge of circularity seem plausible. Laudan, Shapere, and others have made very detailed arguments against the Kuhnian picture that all standards are internal to a theory. Furthermore, as Ian Hacking and Peter Galison have emphasized, 'experimentation has a life of its own' and often follows constraints that are independent of the theories under consideration. Peter Kosso points out that even in cases of very indirect observations, which are heavily theory-laden, it is quite possible for the theory under test to be epistemically independent of the auxiliary theories that are used to construct the observing instrument. Once we reject the view that all standards are internal to a theory, we see that it is an open question whether a given theory is related to another epistemically.

In summary, although I follow Friedman very closely in his development of the constitutive elements of scientific theory, and I agree with his critique of Quine that holism masks a part of the structure of science to which we should be paying close attention, I believe that Friedman overreaches with his notion of a philosophical meta-paradigms or meta-frameworks. Indeed, it seems clear that Friedman wants something to replace what we had in the Kantian a priori, something that stands outside of science and that is universal and transhistorical, at least at the ideal end of inquiry.[3] My view is that science has the resources to make objective judgments without an outside arbiter and further that there are no universal and transhistorical principles. I take one of the lessons of pragmatism to be that we can get along just fine without any universal and fixed principles. Living without absolutes does not lead to relativism or social construction, but it does define the human condition.

HACKING'S STYLES OF REASONING

Ian Hacking has been writing about conceptual change and the history of scientific ideas for a long time. His term to describe science and scientific change is 'style of reasoning,' which he took over and adapted from A. C. Crombie. Like Kuhnian paradigms, these describe a science in terms of a framework that makes it possible. Hacking now says that

> '[s]tyles of scientific thinking and doing' is a better label; the styles can also be called genres, or, ways of finding out. . . . Ways of finding out are not defined by necessary and sufficient conditions, but can be recognized as distinct within a sweeping, anthropological, vision of the European sciences. The approach is unabashedly whiggish. The emergence of these styles is part of what Reviel Netz calls cognitive history, and is to be understood in an ecological way. How did a species like ours, on an Earth like this, develop a few quite general strategies for finding out about, and altering, its world? (2012, 599)

There are at least two explicit links between Hacking's styles of reasoning and the theories of the constitutive elements that I have been describing here. First, he puts his project in Kantian terms. Like all the theories of the constitutive elements in science, Hacking is describing the conditions of the possibility of science and in particular of the possibility of objectivity:

> My study is a continuation of Kant's project of explaining how objectivity is possible. He proposed preconditions for the string of sensations to become objective experience. He also wrote much about science, but only after his day was it grasped how communal an activity is the growth of knowledge. Kant did not think of scientific reason as a historical and collective product. We do. My styles of reasoning, eminently public, are part of what we need to understand what we mean by objectivity. (Hacking [1992] 2002, 181)

Of course, unlike Kant, the framework is not fixed for all time but, rather, changes in conceptual revolutions. The second point of connection between styles of reasoning and the theories of the constitutive elements in science is this change, which Hacking calls the historical a priori:

> The historical *a priori* points at conditions on the possibilities of knowledge within a "discursive formation;" conditions whose dominion is as inexorable, there and then, as Kant's synthetic *a priori*. Yet they are at the same time conditioned and formed in history, and can be uprooted by later, radical, historical transformations. T. S. Kuhn's paradigms have some of the character of a historical *a priori*. (Hacking 2002, 5)

We see that Hacking's approach uses the same terminology of a changing a priori that we have seen in other theories of the constitutive elements in science. The a priori gains a prefix, whether it be pragmatic, functional, dynamic, or historical, and is thus changed dramatically. However, Hacking goes on to reject putting his "historical ontology" in terms of an a priori, changing or not:

> For the nonce, I think that philosophy in the twentieth century drank its fill at the Kantian source, and should now turn back to more empirical springs. . . . There is plenty of room for history plus philosophy without reincarnating the synthetic *a priori* in historicist garb. (Hacking 2002, 5)

As I have noted before, I think that holding onto the term *a priori* even with an adjective in front of it has been seriously misleading, so I agree with Hacking's rejection of the term. I agree as well that we should stay closer to empiricism than to Kantianism. Since I am describing theories of constitutive elements, however, not theories of the a priori, I think that Hacking's styles of reasoning can be put into my category. Hacking introduces the idea of styles of reasoning in "Language, Truth, and Reason" ([1982] 2002) and elaborates on the ideas in many more recent pieces, most notably "'Style' for Historians and Philosophers" ([1992] 2002). As Martin Kusch notes in an extensive critical analysis, one of the main changes in Hacking's presentation of styles of reasoning over the years is that the early article seems endorse a kind of relativism while in his later writings, Hacking carefully avoids relativism (Kusch 2010). In fact, I think that a case can be made that even in the early article, Hacking is careful to avoid certain forms of relativism. In particular, he is at pains to show that his view is not subject to the anti-relativist argument that Donald Davidson mounts in "On the Very Idea of a Conceptual Scheme" (Davidson 1973; Hacking [1982] 2002, 172–174). That said, Hacking is very aware of the challenge that is faced by the charge of relativism. While Kusch seems quite willing to adopt relativism, Hacking clearly wants to distance himself from it. To a certain extent, their dispute could simply be a matter of rhetoric: What language should you use to express the idea of historical ontology (Hacking) or historical epistemology (Kusch)? Should we call these positions relativist or maintain that they hold onto a notion of objectivity? I side with Hacking here and think that the rhetoric of relativism is to be avoided, largely because it is misleading and leads to irrelevant critiques. Nevertheless, I am not an antirelativism crusader like many philosophers of science. In fact, I think that most of the time, the cure offered for relativism is worse than the disease. We should reject all absolutes and start from the premise that everything is historical, contingent, and situated.

Unlike Kusch, however, I think that Hacking does have the tools to block the charge of relativism. Kusch takes up this issue of relativism in section 3.2 of his detailed and insightful critical analysis of Hacking's historical ontology (2010, 166ff.). I agree with many of the points raised by Kusch, but right from the start, I want to disagree with the way Kusch characterizes relativism. On one hand, this may make it seem that I am merely raising a verbal issue, what we should call relativist, rather than the merits of Kusch's view. Indeed, I think that part of the dispute is exactly such positioning and in that sense verbal, but I also think that there are also good reasons for rejecting the label 'relativism.' Kusch starts with a definition of 'epistemic ambivalence':

> Fundamental disagreements over the rationality and justification of beliefs can motivate a reaction of 'epistemic ambivalence': we recognize that our interlocutor on the other side has—seen from her perspective—perfectly legitimate and rational reasons for her judgements, and we appreciate that we can argue for the superiority of our position only by begging the question against her. This does not mean that we abandon our own judgements, but it means that we come to see them in a new light: as relative to our epistemic system. (2010, 167)

So far, I would claim, all that we have is a polite fallibilism, not relativism. I can surely think that my interlocutor is rational but wrong, while recognizing the fallibility of my own judgments and the fallibility of my own methods of argument. I can also recognize the situated nature of my own judgments, but there is no reason that I should be led to think that opposing positions are equally valid, which I take to be the hallmark of relativism.[4] As for the idea that my views are (merely) relative to my own epistemic system, I would say for Hacking, the answer is yes in one sense and no in another. Of course, it is true that my judgments can only be formulated in my style of reasoning, but once that style is adopted, the grounds of the judgment is objective—in pragmatist terms: it either works or it does not, and whether it does is not built into the style of reasoning in advance.

The key point is that Kusch does not take seriously enough Hacking's insistence that styles of reasoning tell us what is up for grabs as true or false, rather than what is true.[5] Styles would determine what is true in an axiomatic purely deductive system, such as mathematics, but not in the empirical sciences. In mathematics, once you have set out your axioms and definitions, everything that can be determined has been. You may spend years trying to figure out all of the consequences of the system that you have defined and some questions may turn out to be undecideable, but in a strong sense everything is determined in advance, even whether or not something is decidable. This is quite a striking contrast with an empirical science, where setting out the basic laws and definitions does not determine what is true. You have to go out and do things—experiment, build things, intervene in nature—in order to find out what is true. Hacking's point is that in the empirical sciences all that the style of reason tells you is what is a possible candidate for truth or falsity, not what is true or false. The style gives you the words, the concepts, and even the physical practices that you need in order to express yourself and intervene in the world.

In some sense, the style limits us, but in a much stronger sense, it enables us to carry out the practice of science. We might take this point in analogy with language itself. In order to be understood, we have to use words with the standard meanings and follow the rules of grammar (more or less). Yes, this not only limits what we can say but also enables us to be understood by others. Consider Kusch's critique of Hacking's argument against social constructivism, which appears in Hacking's work on contingency of science (Hacking 1999, 2000):

> in work not directly focused on styles of reasoning, Hacking has proposed a more general argument against relativistic social constructivism: while different communities might differ in the kinds of questions they ask and regard as 'live', once their respective questions are 'well-asked', the actual answers are fixed, and not socially constructed (Hacking, 2000, p. 569). I agree to the extent that of course it makes sense to say that different (scientific) traditions often seek to answer different questions, and that the questions of one tradition need not make sense to the members of other traditions. We might also go further and say that the members of scientific traditions usually assume that their 'real' questions have determinate answers. But this is where my agreement ends. These correct and trivial observations do not license the further claim that such questions really do have determinate answers as such, and that these answers are outside the realm of negotiations, interests, contingency and history. (2010, 168)

I do not read Hacking as saying that the answers are outside of the realm of negotiations, interests, contingency, and history. I do read him as saying that none of these things determine an answer, which is what is required to make social construction go through. Instead, Hacking says that aspects of the world determine the answer:

> when the question is a live one, and there is a context in which there are ways of addressing the question, or even methods of verification for

possible answers, then aspects of the world determine what the answer is, even though only people in a scientific society find out the answer. (2000, S69)

Aspects of the world determine what the answer is; the answer is not built into the style in advance. Of course, the answer will still be discovered by a concrete individual at a particular time and place, and negotiations, interests, contingency, and history all come into play, but that is not enough to label all discoveries social constructions.

CONCLUSION

Theories of the constitutive elements in science can be formulated without falling into relativism, idealism or social construction. Kant thought that we needed a fixed and absolute a priori to guarantee objectivity, but we have any such thing. In his later writings, Thomas Kuhn acknowledged the similarity of his view to that of Kant but with a changeable a priori. Michael Friedman has taken up Reichenbach's early position and defended a dynamic a priori, while Ian Hacking has espoused a similar view with his styles of reasoning. In each case, these authors are centrally concerned with conceptual change in science and the philosophical significance of scientific revolutions in the history of science. I argue against Friedman's idea that philosophy plays the role of providing continuity during conceptual change in science, pointing out that nonfoundationalist accounts of science already have the resources to account for scientific revolutions. We do not need a replacement for the a priori in order to maintain objectivity. Yes, all judgments are situated in time and place and are connected to both a concrete individual and a social group, but these judgments can still be objective. Some judgments are biased, some are insufficiently justified, but nevertheless, some are sufficiently justified. The full range of epistemic virtues and faults are possible in judgments that are situated. Calling all situated judgments relative flattens out the epistemic standing of all of our judgments, as if they were all equal to each other, when in fact some are more and some are less justified.

NOTES

1. For a further elaboration of the issue of naturalism, see McArthur (2008) and Klein (2010).
2. See Uebel (2012, 14) for a similar point that such principles are unnecessary and Richardson (2010, 285) for a critique of Friedman's view of rationality.
3. Friedman turns to the work of Habermas for ideas of how to make a quasi-universal philosophical framework for science. Kindi makes a compelling argument that this will not succeed in providing a grounding for the rationality of scientific revolutions (2011, 343).

4. Critics of relativism typically take the claim of equal validity to be central; see, for example, Boghossian (2006). Kusch denies that relativism includes this claim and points to the literature on ethical relativism for support, so it seems that the term *relativism* is used differently in different areas of philosophy.

5. See Elwick (2012, 621) for a somewhat similar defense of Hacking from Kusch's critique.

BIBLIOGRAPHY

Boghossian, Paul. 2006. *Fear of Knowledge: Against Relativism and Constructivism*. Oxford: Oxford University Press.

Chang, Hasok. 2008. "Contingent Transcendental Arguments for Metaphysical Principles." In *Kant and Philosophy of Science Today*, edited by M. Massimi. Cambridge: Cambridge University Press, 113–133.

Davidson, Donald. 1973. "On the Very Idea of a Conceptual Scheme." *Proceedings and Addresses of the American Philosophical Association* 47: 5–20.

Doppelt, Gerald. 1978. "Kuhn's Epistemological Relativism: An Interpretation and Defense." *Inquiry* 21: 33–86.

Elwick, James. 2012. "Layered History: Styles of Reasoning as Stratified Conditions of Possibility." *Studies in History and Philosophy of Science Part A* 43 (4): 619–627.

Friedman, Michael. 2001. *Dynamics of Reason: The 1999 Kant Lectures at Stanford University*. Stanford, CA: CSLI Publications.

———. 2002. "Kant, Kuhn, and the Rationality of Science." *Philosophy of Science* 69 (2): 171–190.

———. 2003. "Kuhn and Logical Empiricism." In *Thomas Kuhn*, edited by T. Nickles. Cambridge: Cambridge University Press, 19–44.

———. 2004. "Philosophy as Dynamic Reason: The Idea of a Scientific Philosophy." In *What Philosophy Is: Contemporary Philosophy in Action*, edited by H. Carel and D. Gamez. London: Continuum, 73–96.

———. 2005a. "Ernst Cassirer and Contemporary Philosophy of Science." *Angelaki: Journal of the Theoretical Humanities* 10 (1): 119–128.

———. 2005b. "Transcendental Philosophy and Twentieth Century Physics." *Philosophy Today* 49 (5 [Suppl.]): 23–29.

———. 2006. "Carnap and Quine: Twentieth-century Echoes of Kant and Hume." *Philosophical Topics* 34: 35–58.

———. 2008a. "Einstein, Kant and the A Priori." In *Kant and Philosophy of Science Today*, edited by M. Massimi. Cambridge: Cambridge University Press, 95–112.

———. 2008b. "Ernst Cassirer and Thomas Kuhn: The Neo-Kantian tradition in history and philosophy of science." *Philosophical Forum* 39 (2): 239–252.

———. 2009. "Einstein, Kant and the Relativized A Priori." In *Constituting Objectivity: Transcendental Perspectives on Modern Physics*, edited by M. Bitbol, P. Kerszberg, and J. Petitot. Berlin and New York: Springer, 253–267.

———. 2010a. "A Post-Kuhnian Approach to the History and Philosophy of Science." *The Monist* 93 (4): 497–517.

———. 2010b. "Synthetic History Reconsidered." In *Discourse on a New Method: Reinvigorating the Marriage of History and Philosophy of Science*, edited by M. Domski and M. Dickson. Chicago and La Salle, IL: Open Court, 571–813.

———. 2011. "Extending the Dynamics of Reason." *Erkenntnis* 75 (3): 431–444.

———. 2012. "Reconsidering the Dynamics of Reason: Response to Ferrari, Mormann, Nordmann, and Uebel." *Studies in History and Philosophy of Science Part A* 43 (1): 47–53.

Galison, Peter. 1987. *How Experiments End*. Chicago: Chicago University.

———. 1988. "History, Philosophy, and the Central Metaphor." *Science in Context* 2: 197–212.

————. 1989. "Multiple Constraints, Simultaneous Solutions." In *PSA 1988, vol. 2.* East Lansing, MI: Philosophy of Science Association, 157–163.

Hacking, Ian. (1982) 2002. "Language, Truth, and Reason." In *Rationality and Relativism*, edited by M. Hollis and S. Lukes. Oxford: Basil Blackwell, 48–66. Rpt. *Historical Ontology*, Cambridge, MA: Harvard University, 159–177.

————. 1983. *Representing and Intervening.* New York: Cambridge University Press.

————. (1992) 2002. "'Style' for Historians and Philosophers." *Studies in History and Philosophy of Science* 23: 1–20. Rpt. *Historical Ontology*, Cambridge, MA: Harvard University, 178–199.

————. 1999. *The Social Construction of What?* Cambridge, MA: Harvard University Press.

————. 2000. "How Inevitable Are the Results of Successful Science?" *Philosophy of Science* 67: S58–S71.

————. 2002. *Historical Ontology.* Cambridge, MA: Harvard University Press.

————. 2012. "'Language, Truth and Reason' 30 Years Later." *Studies in History and Philosophy of Science Part A* 43 (4): 599–609.

Kindi, Vasso. 2011. "The Challenge of Scientific Revolutions: Van Fraassen's and Friedman's Responses." *International Studies in the Philosophy of Science* 25 (4): 327–349.

Klein, Alexander. 2010. "Divide et Impera! William James and Naturalistic Philosophy of Science." *Philosophical Topics* 36 (1): 129–166.

Kosso, Peter. 1988. "Dimensions of Observability." *British Journal of Philosophy of Science* 39: 449–467.

————. 1989. *Observability and Observation is Physical Science.* Dordrecht: Kluwer.

Kuhn, Thomas S. 1962. *The Structure of Scientific Revolutions.* Chicago: University of Chicago Press.

————. 1993. "Afterwords." In *World Changes*, edited by P. Horwich. Cambridge, MA: MIT Press, 311–341.

————. 2000. *The Road since Structure.* Chicago: University of Chicago Press.

Kusch, Martin. 2010. "Hacking's Historical Epistemology: A Critique of Styles of Reasoning." *Studies in History and Philosophy of Science Part A* 41 (2): 158–173.

Laudan, Larry. 1977. *Progress and its Problems: Towards a Theory of Scientific Growth.* Berkeley: University of California Press.

————. 1987. "Relativism, Naturalism and Reticulation." *Synthese* 71: 221–234.

McArthur, Dan. 2008. "Structural Realism, the A Priori and Normative Naturalism." *International Studies in the Philosophy of Science* 22 (1): 5–20.

Richardson, Alan. 2010. "Ernst Cassirer and Michael Friedman: Kantian or Hegelian Dynamics of Reason?" In *Discourse on a New Method: Reinvigorating the Marriage of History and Philosophy of Science*, edited by M. Domski and M. Dickson. Chicago and La Salle, IL: Open Court, 279–294.

Rouse, Joseph. 2003. "Kuhn's Philosophy of Scientific Practice." In *Thomas Kuhn*, edited by T. Nickles. Cambridge: Cambridge University Press, 101–121.

Shapere, Dudley. 1980. "The Character of Scientific Change." In *Scientific Discovery, Logic and Rationality*, edited by T. Nickles. Dordrecht: D. Reidel, 648–649.

————. 1984. *Reason and the Search for Knowledge.* Dordrecht: D. Reidel.

————. 1986. "Objectivity, Rationality, and Scientific Change." In *Proceedings of the Biennial Meeting of the Philosophy of Science Assoxiation 1984*, vol. 2, edited by P. Kitcher and P. Asquith. East Lansing, MI: Philosophy of Science Association, 637–663.

Stump, David J. 1991. "Fallibilism, Naturalism and the Traditional Requirements for Knowledge." *Studies in History and Philosophy of Science* 22: 451–469.

Uebel, Thomas. 2012. "De-Synthesizing the Relative A Priori." *Studies in History and Philosophy of Science Part A* 43 (1): 7–17.

8 On the Role of Mathematics in Physical Theory

Mathematics plays a prominent role in the theories of the constitutive elements of science. Mathematics seems to have a special status, appearing to be a priori and yet being useful in empirical theory. The indispensability argument in the philosophy of mathematics is used by some to argue for Platonism by claiming that the use of abstract entities of mathematics in successful empirical theory can only be explained by the existence of these abstract entities. Part of this argument is in keeping with the argument for constitutive elements in science because mathematics is seen as a necessary precondition for further inquiry, but the confirmational holism that the indispensability argument makes use of can be rejected by the theories of the constitutive elements in science. I briefly survey the literature on the indispensability argument and argue that it is independent of the issue of the constitutive elements in science. The applicability of mathematics to the physical world and its use in empirical science can be explained in several ways that are compatible with the existence and functioning of a constitutive element in science, so even though mathematics is the most striking and persistent example of the constitutive element in science, highlighting its constitutive role does not commit one to Platonism. However, the question of why mathematics can be so successfully used in explaining physical phenomena remains puzzling.

As we have seen throughout the book, mathematics plays a special role in science. Furthermore, it may play a special role because it is always taken to be constitutive (and a priori), that is, that the basic epistemic status of mathematics never seems to change from one theory to another, throughout all of the revolutions in science. I want to argue that like all of the constitutive elements in science, applied mathematics does not seem to be easily fit into categories as either analytic or synthetic, or as empirical or a priori. The issues concerning the status of mathematics can be brought into focus by considering the indispensability argument in the philosophy of mathematics. I would like to offer a pragmatic solution to the problem of the indispensability of mathematics, at least insofar as it pertains to the special role that mathematics plays in the empirical sciences. Some may call this a pragmatic nonanswer to the problem, but that is what you get with deflationary strategies. At least the pragmatist can offer a diagnosis of how the problem arises and make suggestions as to how we should answer the question. My suggestion is that there will be no general answer to the question of why mathematics can be successfully applied in science. Furthermore, answering the question does not require a particular stance in the philosophy of mathematics such as Platonism. There are many strategies for explaining this success, and besides, settling the issues in the philosophy of mathematics is outside of the scope of this work, and I do not take a stand on those issues here.

THE INDISPENSABILITY ARGUMENT

Mark Colyvan begins his recent overview and analysis of the indispensability argument as follows:

> From the rather remarkable but seemingly uncontroversial fact that mathematics is indispensable to science, some philosophers have drawn serious metaphysical conclusions. In particular, Quine (1953; 1976; 1980; 1981a; 1981c) and Putnam (1979a; 1979b) have argued that the indispensability of mathematics to empirical science gives us good reason to believe in the existence of mathematical entities. According to this line of argument, reference to (or quantification over) mathematical entities such as sets, numbers, functions and such is indispensable to our best scientific theories, and so we ought to be committed to the existence of these mathematical entities. To do otherwise is to be guilty of what Putnam has called "intellectual dishonesty" (Putnam 1979b, p. 347). Moreover, mathematical entities are seen to be on an epistemic par with the other theoretical entities of science, since belief in the existence of the former is justified by the same evidence that confirms the theory as a whole (and hence belief in the latter). This argument is known as the Quine-Putnam indispensability argument for mathematical realism. (2011, 1)

There are two parts to the argument. First, it is argued that the general argument for scientific realism applies to mathematics as well. Given that our successful theories mention various mathematical entities, it is argued that our theories could not be successful (short of a miracle) unless those entities actually existed. We could of course stop right here by questioning the argument for scientific realism. If the argument for scientific realism does not go through, then a fortiori, the indispensability argument for mathematical realism does not go through.[1] So there are already options for someone who wished to question the argument, but let us continue to play along and accept the argument for scientific realism. Furthermore, there is the assumption that mathematics has to be taken literally; that is, we need to reject fictionalist accounts of mathematics.[2] Thus, again, there are controversial issues that leave open a place to question the indispensability argument, but again I continue because the next step is the most relevant to the discussion of the constitutive elements in science.

The indispensability argument further claims that mathematics is justified because it is used in successful science, that is, that mathematics receives confirmation by the fact that it is used in successful science. This is an extension of Quine's confirmational holism to mathematics. Anyone holding that mathematics is justified a priori (and only a priori) will reject this extension of Quine and, so to speak, take mathematics off the table. Here is Colyvan's explanation of this part of the indispensability argument:

> Confirmational holism is the view that theories are confirmed or disconfirmed as wholes (Quine 1953, p. 41). So, if a theory is *confirmed* by empirical findings, the *whole* theory is confirmed. In particular, whatever mathematics is made use of in the theory is also confirmed (Quine 1976, pp. 120–122). Furthermore, it is the same evidence that is appealed to in justifying belief in the mathematical components of the theory that is appealed to in justifying the empirical portion of the theory (if indeed the empirical can be separated from the mathematical at all). (2011, 6)

To recapitulate, there are several issues here that need to be separated and there are two major steps in the indispensability argument. First, it is argued that theories need to be taken literally or on face value. If mathematical theories mention sets, numbers, points, and so on, then even though these are special abstract entities, they are to be taken to exist and mathematical statements about them are true (or false) in a straightforward sense. Second, Quine's confirmational holism is assumed; that is, it is taken for granted that the successful application of mathematics justifies it.

Theories of the constitutive elements in science would most naturally be seen as rejecting Quine's confirmational holism and holding that mathematics is sui generis and a priori. Quineans would say that what was once called a priori is simply the hard core of our empirical theories, that is, the most well-established parts of our physical theory. However, the constitutive principles are not necessarily more entrenched and frequently they have never been *directly* tested. As noted earlier, Friedman points out that the difference between the constitutive and the nonconstitutive is not a difference in the degree of justification (Friedman 2001, 46). Indeed the notion of degree of justification will not at all do justice to the difference between the constitutive principles of physical science and the empirical ones. Quine would have us believe that the core is entrenched; that is, it consists of statements that have been around for a long time and that are accepted by everyone. Neither of these two conditions is necessarily met by the constitutive elements. For example, Newton proposed the calculus at essentially the same time as his physics, but calculus is clearly a necessary precondition to working in Newtonian physics, so it counts as constitutive in that context. The calculus was not entrenched in any sense at that time, because it was brand new and it was controversial. The calculus did not become uncontroversial until after the work of Cauchy, Weierstrass, and others put it on a sound footing. The rejection of Quinean confirmational holism does not necessarily, however, have a direct impact on the first part of the indispensability argument. Is there still an analogy with scientific realism about the existence of mathematical objects?

OBJECTIONS TO THE INDISPENSABILITY ARGUMENT

Before coming back to considering other options for constitutive theories, let us consider a couple of the major objections to the indispensability argument extant in the literature to see how they have dealt with the issue. Elliot Sober (1993) raises a strong objection to the confirmational holism that Quine advocates. Sober points out that it is simply not the case that mathematics receives justification by its applications in science. Mathematics receives (pure) mathematical justification, not empirical justification. Furthermore, if we obtain negative data, we never blame the mathematics, as if we could falsify it by empirical testing (Sober 1993, 49–50). Sober puts his argument against the indispensability argument in the context of the debate over scientific realism. Scientific realists claim that they have arguments for the existence of theoretical entities and for the existence of mathematical entities. Bas van Fraassen denies that the arguments for either go through. Sober contends that the realist's argument for the existence of theoretical entities may have merit, but not the argument for mathematical entities. The reason, he says, is that all competing hypotheses equally assume the mathematics, so the mathematics is not really being tested (Sober 1993, 44–45). Indeed, it is hard to imagine how it could be tested.

How has Sober argued against the first part of the indispensability argument, the part that leads to the existence of mathematical entities? Even his contrastive empiricism relies on arguing that mathematics does not receive empirical confirmation. So in other words, Sober rejects the idea that there is a parallel to scientific realism in the argument to be made about mathematical objects. This seems to me to be a surprisingly weak argument. Sober makes a very strong case against confirmational holism, but he has done little more than simply reject the parallel to scientific realism. This is a version of Resnik's response to Sober (Resnik 1995, esp. 170), which I consider in the following.

Like Sober, Susan Vineberg argues that the extension of scientific realism to mathematical entities is a nonsequitur. Mathematics does not receive confirmation from our best physical theories, she argues, despite the fact that it plays an indispensable role in those theories. She provides a diagnosis of why this is the case, namely that the indispensable objects in physical science are causally connected to our observations, but the mathematical entities are not:

> The situation is different for mathematical entities. These apparently play no causal role in our scientific theories. Hence, the reasons we have for understanding scientific claims about theoretical entities literally do not similarly provide a reason to read the mathematics used in science literally. Quine and Putnam held that our scientific theories as a whole are to be taken literally and that, so understood, we have good evidence that our best scientific theories are true. Consequently, they maintained that the mathematics that occurs in these theories is confirmed as literally true. I have argued that the evidence that supports taking claims about the causal properties of theoretical entities as literally true does not similarly support the claim that the mathematics that occurs in science is literally true. (Vineberg 1996, S261)

Sober and Vineberg argue against the indispensability argument for the existence of mathematical entities by rejecting the parallel with the scientific realist argument for the existence of unobservables. Sober thinks that by rejecting confirmational holism he has blocked the move made by advocates of the indispensability argument, and Vineberg underscores the fact that mathematical objects are different from unobservables in physical theory. But what if someone were to simply insist that causal relations are unnecessary for the argument, that anything mentioned in a theory deserves some commitment to its existence? And what if someone were to argue that mere reference is enough to justify an existence claim, independently of whether the mathematical parts of our theories are confirmed by empirical results? Confirmational holism seems to be a further claim, beyond the argument that parallels the argument found in scientific realism. So I think we need something more than the denial of confirmational holism to refute the indispensability argument. Of course, as I mentioned earlier, one way to do this is to reject the realist form of argument altogether or to adopt fictionalism about mathematics, but I do not think that theories of the constitutive elements in science need to be linked to a rejection of realism or to fictionalism. The issues are independent and one could be a scientific realist or antirealist who accepts the idea that there are constitutive elements in science. Similarly, one could hold various views in the philosophy of mathematics and still accept the idea as a framework for understanding science.

As I have mentioned, it is the thesis of confirmational holism that is most obviously in conflict with theories of the constitutive elements in science. I am a little hesitant about making mathematics totally independent of the rest of science, however, for I have argued throughout the book that there are no absolute distinctions, only continua, so we cannot draw a hard and permanent distinction between what is empirical and what is not. Denying absolute distinctions makes it a little harder to apply Sober's argument, but not much. Mathematics is clearly at the far end of the spectrum, as nonempirical as we can get, given that we do not, and we need not, use empirical observations to justify mathematics. Indeed, Sober does not absolutely rule out empirical confirmation of mathematical ideas but, rather, the notion that empirical results give a general grounding for mathematics (1993, 51). It is also important to note that he does not replace the justification of mathematics with anything, for example not by a priori intuition.

Moving on to consider another major critique of the indispensability argument, I am very sympathetic with Penelope Maddy (1992), who argues that Quine's naturalism and his confirmational holism are in conflict and that we should side with naturalism. Important features of Quine's (1981b) naturalism include the ideas that the philosophy is continuous with science and especially that philosophy is not a ground of the sciences. Maddy's idea is that naturalism, in part, requires us to look at how mathematics is actually practiced and how it is actually used in science. She develops her view of naturalism considerably in recent work (Maddy 1997, 2001, 2007). When we do look at mathematical and scientific practice, we find that mathematics is not justified by its applications but rather by pure mathematics. We cannot use philosophical arguments to make a priori pronouncements about how mathematics is or should be used in science.

Raising another argument against the indispensability argument, Maddy points out that there are all kinds of false theories that are useful in science, so we cannot make sense of the realist claim that the indispensability of mathematics in successful empirical theories shows that the mathematics is true and that mathematical entities really exist:

> [A] glance at any freshman physics text will disappoint this notion. Its pages are littered with applications of mathematics that are expressly understood not to be literally true: e.g., the analysis of water waves by assuming the water to be infinitely deep or the treatment of matter as continuous in fluid dynamics or the representation of energy as a continuously varying quantity. Notice that this merely useful mathematics is still indispensable; without these (false) assumptions, the theory becomes unworkable. (Maddy 1992, 281)

The realism assumed by the indispensability argument is therefore specifically called into question.

Maddy presents what she calls the simple indispensability argument for the existence of mathematical entities and argues that it will run afoul of her version of naturalism: "Simple indispensability will not do, if we are to remain faithful to mathematical practice" (1992, 279). She then presents and defends a revised version of the indispensability argument that she says in closer to mathematical practice, by basically arguing that the indispensability of mathematics only has an indirect relation to foundational parts of mathematics:

> From this point of view, a modified indispensability argument first guarantees that mathematics has a proper ontology, then endorses (in a tentative, naturalistic spirit) its actual methods for investigating that ontology. For example, the calculus is indispensable in physics; the set-theoretic continuum provides our best account of the calculus; indispensability thus justifies our belief in the set-theoretic continuum, and so, in the set-theoretic methods that generate it; examined and extended in mathematically justifiable ways, this yields Zermelo-Fraenkel set theory. (Maddy 1992, 280)

And "[i]n short, legitimate choice of method in the foundations of set theory does not seem to depend on physical facts in the way indispensability theory requires" (Maddy 1992, 289). Thus, foundational parts of mathematics remain isolated from physical theory, but they are connected to parts of mathematics that are required in physical theory. I would go farther than Maddy in taking the position that there is not going to be any general story about how mathematics is used in science, that rather we must look case by case at the role that mathematics is playing. I think those looking at mathematical practice in detail are on the right track and that we should not spend our time arguing about general metaphysical issues about the existence or nonexistence of mathematical entities (see, e.g., Batterman 2006, 2010; Pincock 2007, 2012; Mancosu 2008).

RESNIK'S DEFENSE

We should consider a defense of the indispensability argument as well and the most nuanced comes from Michael Resnik (1995, 1997). Resnik specifically addresses Sober and Maddy's arguments and claims to have found ways around their objections. In particular, he makes an argument that is independent of confirmational holism, which, as I noted earlier is that part of the argument to which those advocating a constitutive theory object. Resnik's work on the philosophy of mathematics provides a way to think about the status of mathematics within a modified Quinean perspective. Resnik's main new additions to discussions in the philosophy of mathematics are a deflationary account of reference and truth—a reduced and subtle kind of realism in mathematics—and structuralism to account for the incompleteness of mathematical objects. Resnik's immanent account of reference and truth makes for realism that some may consider incomplete, but clearly deflationary accounts are not antirealist, since the epistemic accounts of truth usually associated with antirealism are rejected. Of particular significance is the point that an immanent account of reference eliminates one of the strongest arguments against mathematical realism. A realist approach seems to require a causal theory of reference, and since mathematical objects are causally inert, it seems impossible to be a metaphysical realist in mathematics. This is one of the lessons drawn from Benacerraf's (1983) famous argument about the unknowability of mathematical objects, given the fact that we do not interact with them causally.

By advocating structuralism, Resnik hopes to make the ontological and referential relativity of mathematics seem natural. Since mathematical objects are defined by the axioms and only up to isomorphism, there is an apparent problem for the mathematical realist, given that typically multiple objects satisfy the axioms since there are infinitely many structurally identical models. Resnik argues that realism can account for the incompleteness of mathematical objects as well as other philosophies of mathematics do, so that he can argue that realism has other advantages as a philosophy of mathematics (1997, 92). His aim is to clarify the nature of mathematical objects, not to reduce them to one kind of object (Resnik 1997, 223). Not only is structuralism intended to make the incompleteness of mathematical objects philosophically palatable; it is also taken to be compatible with Resnik's epistemology, or at least with his story of the genesis of mathematical knowledge.

Resnik defends the indispensability thesis and in his book (1997) promotes Quinean holism, arguing that mathematics is not separable from the rest of science and that the justification of mathematics, once we move away from local justifications within various branches of mathematics, ultimately involves the role that mathematics plays in science as a whole. As I mentioned previously, I am more comfortable giving explanations at the local level. Resnik defends epistemic holism by arguing that the objects of physics are not so different from those in mathematics and by accounting for the stability of mathematics pragmatically. Mathematics will be tinkered with less frequently than experimental hypotheses, becoming relatively a priori, that is, more likely to be taken for granted. However, Resnik admits that mathematics is never falsified (1997, 133), leading one to wonder whether this is not in itself prima facie evidence for a fundamental epistemic difference between mathematics and the rest of science, perhaps enough of a difference for theories of the constitutive element in science. I do not think that rejection of confirmational holism alone undermines the indispensability argument, but Resnik's defense of it and of the Quinean project in general does not seem to be strong enough to undermine the theories of the constitutive elements in science.

Let us consider again how the indispensability argument is supposed to be working. It is not only an argument for the existence of mathematical objects; it is also supposed to be an answer to the question of how mathematics can be so successfully used in the sciences. The argument is parallel to the argument for realism in physical science. Realism in physical science is supposed to be explanatory by claiming that successful theories work because they describe reality as it actually is. By analogy, the existence of mathematical objects would explain why mathematics works, because it correctly describes features of actual things, such as numbers and sets. That explains the success of pure mathematics, but it still leaves open the question of why mathematics can be successfully applied to physical objects.[3] What do abstract mathematical entities and physical objects have to do with each other? Benacerraf's famous question about how mathematical entities can be known, given that no causal relation with them is possible, remains unanswered, and his question can be extended to that of how mathematical objects can relate to physical things. There are many possible answers to the question, but the indispensability argument in itself does not answer the question. Furthermore, if we grant that sometimes mathematics works and sometimes it does not, then the question raised by the indispensability argument seems to be why the mathematics that works well works so well, a question that may seem to be approaching a tautology.

I contend that the indispensability argument does not give us any answer to the question of how mathematics can be successfully used in science. If the puzzle is why mathematics is successful, then, parallel to the realism argument in physical science, we get the answer that it is successful because the things that are described in mathematics actually exist, just as the things described in physical science actually exist. However, there is a major gap left open in this explanation, namely, the relation between the abstract entities in mathematics and the concrete physical objects in physical science. The parallel between realism in physical science and the indispensability argument for mathematical Platonism, even if we grant both, leaves the major question of why mathematics can be used successfully in science unanswered. Here is a nice, concise statement of the issue of how mathematics can be explanatory:

> Mathematics seems to have a subject matter that is distinct from the physical world. The problem, then, is to say what connection there is between the physical world and mathematics that can explain the successful application of mathematics in scientific reasoning. (Pincock 2004, 137)

The traditional answer to this problem was of course to say that mathematics is a priori in a full-bodied sense. According to Kant, mathematics is synthetic; that is, it tells us substantial things about the way the world is (at least the phenomenal world), and a priori. In the contemporary literature, I suggest that we consider how mathematics is actually used in science. Chris Pincock gives a good answer to the question of why mathematics can be successfully applied with his analysis of models—there are similar structures in mathematics and in the reality that it represents (2012, 21). In this type of case, Pincock has shown us one way of understanding the successful application of mathematics in physical theory.

MATHEMATICAL PRACTICE

As I mentioned mathematical practice is being explored by many authors but I would like to focus on Pincock's (2012) recent book, in part because he takes up a discussion of the constitutive elements in science. Interestingly, Pincock sees constitutive frameworks as one kind of use of mathematics in science. He surveys ways in which mathematics has been successfully used in scientific representations of phenomena and then turns in his chapter 6 to the use of mathematics in constitutive frameworks, surveying the views of Carnap, Kuhn, and Friedman.

To give a sense of the kind of cases that Pincock considers in his work, his nice example of the Königsberg bridge problem is a good place to start. The problem was how to take a walk and cross each bridge in the city exactly once. Euler showed that it is impossible to do so and that it is mathematics that explains why, not anything to do with the physical structure of the bridges:

> The abstract explanation seems superior because it gets at the root cause of why walking a certain path is impossible by focusing on the abstract structure of system. Even if the bridges were turned into gold, it would still have the structure of the same graph, and so the same abstract explanation would apply. By abstracting away from the microphysics, scientists can often give better explanations of the features of physical systems. (Pincock 2007, 260)

As Pincock points out, the material structure of the bridges is irrelevant in this case; only the locations of the bridges make a difference. Therefore, we have an example of a mathematical explanation of a physical fact, namely, that it is impossible to go for a walk and cross each bridge exactly once. The example shows one kind of role that mathematics can play in scientific representation. Pincock thinks that the use of mathematics as part of constitutive frameworks is another role, but he has some doubts about how they are described by Carnap, Kuhn, and Friedman.

I have noted that Friedman's dynamic a priori denies two important theses of Quine's; that is, it denies both naturalism and holism. Quine is the founder of the indispensability argument and dynamic a priori is formulated in opposition to Quine's views. Nevertheless, there is something parallel in the idea of necessary preconditions and indispensability, and this forms the starting point for Pincock's consideration of theories of the constitutive elements in science. However, some of Pincock's critique of Friedman seems off the mark. When he complains about Friedman's semantic views (Pincock 2012, 131ff.) and in particular the idea that constitutive elements make theories a candidate for truth, it seems to me that Pincock is missing the point. The central point is that, in a practical sense, it is impossible to create a science without certain presuppositions and without certain tools in place. The point is not that Newtonian physics would be neither true nor false without the calculus but, rather, that it could not even be expressed— it could never have been developed—without the calculus. It is not as if Newtonian mechanics existed somehow out there in Platonic heaven, waiting for the calculus to make it true or false. Rather, Newton and those who followed him could not have done their work in physics without developing the calculus first. It is noteworthy that the conditions for the truth or falsity argument are mostly associated with Hacking ([1982] 2002), who introduced it as part of his explanation of styles of reasoning. The argument seems to play a rather minor role in Friedman's account.

There is another argument that I think is more important but also questionable. Pincock argues that mathematics must be a priori in an absolute sense, not just relative to the rest of theory. His argument does not go through, but it is instructive to see how he finds his conclusion. Pincock rightly says that we need the mathematics prior to applying it to physical situations, which is an innocuous statement that does follow from the discussion of the constitutive elements in science. However, he then claims that we need to justify the mathematics prior to using it, which of course leads him to claim that the justification must be a priori, that is, prior to any use in applications. With this argument Pincock seems to ignore the obvious possibility of hypothetical use of the mathematics. We might simply adopt some mathematical technique without any justification (or without knowing the justification), just to try to see if we can make use of it in an application. We do not need justification of the mathematics in advance, no more than we need justification of physical hypotheses in advance. The whole point of the hypothetical-deductive method is precisely to make conjectures and to test them, rather than to justify everything before its use. Pincock is aware of this point and has an argument against it, namely that when a test result is negative, mathematics is never questioned (2012, 218–219). In other words, the Quinean holist is simply wrong to think that the mathematics is on a par with other kinds of assumptions. I am sympathetic to the point against holism, but I do not see how this shows that mathematics must be justified in advance of its use. Pincock would seem to say that there must be a reason that the mathematics is not questioned and that the reason must be prior to the empirical test. It seems to me however that a scientist may grab any mathematical tool off the shelf to see if it is useful, not worrying about the mathematical justification for it at all. Pincock presents many compelling examples of the use of mathematics in science, and he has gone a long way toward explaining why mathematics can be successfully applied to understanding of the physical world, but I do not think that his critique of the theories of the constitutive elements in science is compelling.

CONCLUSION

The overall ground of mathematics is a large and seemingly intractable question: Is mathematical knowledge synthetic a priori, analytic, or empirical? The three great schools in philosophy of mathematics—intuitionist, formalist, and empiricist—line up well with these three positions. I agree with Resnik that the justification of a particular element of mathematics in a given context is justification enough. Pure mathematics will be justified based on consistency and intrinsic interest, whereas applied mathematics will be justified based on its utility in a physical theory. No more is required in order to understand science. I suggest that a general answer to the question of what grounds mathematics is not required for an understanding of how science works. We can look instead to the justification of the mathematical elements of a theory that take place in a scientific context. In the case of pure mathematics, the justification will look more formal and a priori, while in the case of applied mathematics, it may look more utilitarian and empirical.

The question of how mathematics can be applicable to the physical world has been discussed by many philosophers under the rubric of the indispensability argument in the philosophy of mathematics. Mathematics plays a prominent role in the theory of the constitutive elements in science given that mathematics seems to have a special status, appearing to be a priori and yet being useful in empirical science. Furthermore, the epistemological status of mathematics can appear to be fixed; that is, it is always a priori across all scientific revolutions. Surveying the literature on the applicability of mathematics, I conclude that how one answers the indispensability argument does not affect my argument on the role of the constitutive element in science. The applicability of mathematics to the physical world and its use in empirical science can be explained in ways that is compatible with the theory of the constitutive elements in science.

NOTES

1. Resnik (1995, 173) makes a case that mathematics is used in formulating antirealism; therefore, even scientific antirealists are committed to the truth of mathematics and the existence of mathematical entities, but in doing so he assumes that the basic form of argument for realism is valid and that is, of course, denied by nonrealists.
2. See Vineberg (2008) for the issues that remain for nominalists.
3. See Steiner (1989, 1995, 1998) for an extended treatment of the topic.

BIBLIOGRAPHY

Batterman, Robert W. 2006. *The Devil in the Details: Asymptotic Reasoning in Explanation, Reduction, and Emergence.* Oxford: Oxford University Press.

———. 2010. "On the Explanatory Role of Mathematics in Empirical Science." *The British Journal for the Philosophy of Science* 61 (1): 1–25.

Benacerraf, Paul. 1983. "Mathematical Truth." In *Philosophy of Mathematics*, edited by P. Benacerraf and H. Putnam. Cambridge: Cambridge University Press. 403–420.

Colyvan, Mark. 2011. "Indispensability Arguments in the Philosophy of Mathematics." In *The Stanford Encyclopedia of Philosophy* (Spring 2011 edition), edited by E. N. Zalta. http://plato.stanford.edu/archives/spr2011/entries/mathphil-indis/.

Friedman, Michael. 2001. *Dynamics of Reason: The 1999 Kant Lectures at Stanford University.* Stanford, CA: CSLI Publications.

Hacking, Ian. (1982) 2002. "Language, Truth, and Reason." In *Rationality and Relativism*, edited by M. Hollis and S. Lukes. Oxford: Basil Blackwell, 48–66. Rpt. *Historical Ontology*, Cambridge, MA: Harvard University, 159–177.

Maddy, Penelope. 1992. "Indespensibility and Practice." *The Journal of Philosophy* 89 (6): 275–289.

———. 1997. *Naturalism in Mathematics.* Oxford: Clarendon Press Oxford University Press.

———. 2001. "Naturalism: Friends and Foes." *Noûs* 35: 37–67.

———. 2007. *Second Philosophy: A Naturalistic Method.* Oxford: Oxford University Press.

Mancosu, Paolo, ed. 2008. *The Philosophy of Mathematical Practice.* Oxford: Oxford University Press.

Pincock, Christopher. 2004. "A Revealing Flaw in Colyvan's Indispensability Argument." *Philosophy of Science* 71: 61–79.

———. 2007. "A Role for Mathematics in the Physical Sciences." *Nous* 41 (2): 253–275.

———. 2012. *Mathematics and Scientific Representation.* Oxford: Oxford University Press.

Putnam, Hilary. 1979a. "What Is Mathematical Truth." In *Mathematics Matter and Method: Philosophical Papers, Volume 1*, 2nd ed. Cambridge: Cambridge University Press, 60–78.

———. 1979b. "Philosophy of Logic." In *Mathematics Matter and Method: Philosophical Papers, Volume 1*, 2nd ed. Cambridge: Cambridge University Press, 323–357.

Quine, Willard Van Orman. 1953. "Two Dogmas of Empiricism." In *From a Logical Point of View.* Cambridge, MA: Harvard University Press, 20–46.

———. 1976. "Carnap and Logical Truth." In *The Ways of Paradox and Other Essays*, rev. ed. Cambridge, MA: Harvard University Press, 107–132.

———. 1980. "On What There Is." In *From a Logical Point of View*, 2nd ed. Cambridge, MA: Harvard University Press, 1–19.

———. 1981a. "Things and their Place in Theories." In *Theories and Things.* Cambridge, MA: Harvard University Press, 1–23.

———. 1981b. "Five Milestones of Empiricism." In *Theories and Things.* Cambridge, MA: Harvard University Press, 67–72.

———. 1981c. "Success and Limits of Mathematization." In *Theories and Things.* Cambridge, MA: Harvard University Press, 148–155.

Resnik, Michael D. 1995. "Scientific vs. Mathematical Realism: The Indispensability Argument." *Philosophia Mathematica* 3 (2): 166–174.

———. 1997. *Mathematics as a Science of Patterns.* Oxford: Oxford University Press.

Sober, Elliott. 1993. "Indispensability and Mathematics." *The Philosophical Review* **102** (1): 35–57.

Steiner, Mark. 1989. "The Application of Mathematics to Natural Science." *The Journal of Philosophy* **86** (9): 449–480.

———. 1995. "The Applicability of Mathematics." *Philosophia Mathematica* 3 (2): 129–156.

———. 1998. *The Applicability of Mathematics as a Philosophical Problem.* Cambridge, MA: Harvard University Press.

Vineberg, Susan. 1996. "Confirmation and the Indispensability of Mathematics to Science." *Philosophy of Science* **63**: S256–S263.

———. 2008. "Is Indispensability Still a Problem for Fictionalism?" In *Philosophy of Mathematics: Set Theory, Measuring Theories, and Nominalism*, edited by G. Preyer and G. P. Heusenstamm. Frankfurt: Ontos Verlag, 132–143.

9 Epilogue
A Pragmatic Theory of the
Constitutive Elements in Science

Indeed, most everybody will admit [to fallibilism] until he begins to see what is involved in the admission—and then most people will draw back.

—Peirce (1950, 58)

Conceptual change during scientific revolutions is a major issue in general philosophy of science and central to the work of Kuhn and all those who followed him, right up to the current work of Friedman and others, yet the significance of scientific revolutions has remained controversial (Nickles 2014, 1–2). Indeed, broad philosophical models of any aspect of science may be questioned on the grounds that they run roughshod over the history of science, especially the contingent, particular context of a scientific practice so important to understanding any episode in the history of science. Some might argue that we do not need philosophical models of science at all, espousing an antitheory view in which all history is local and ruling out the possibility of generalizing claims about science and its history. However, the idea that history can be philosophy free is naïve, just as is the view that science can be value free. An approach of some kind is necessarily adopted when one studies science, historically or otherwise. As far as the scope of claims about science, there need be no a priori limit to the generality of a claim about science; rather, there are only limits on how far one can generalize given the historical evidence at hand. I advocate for one family of philosophical models of conceptual change in science, referred to here as theories of the constitutive elements in science, that reflects a more accurate image of science than that of competing philosophical models such as Quinean empiricism, and I claim that these models can be used without violating the standards of the historiography of science.

In order to account for conceptual change in science, Friedman revived Reichenbach's idea of a dynamic a priori, showing that conceptual revolutions occur in science when there is a change in what had been taken to be a priori knowledge. The dynamic theory of the (former) a priori understands conceptual change in science as change in fundamental presuppositions that are required for the practice of science, a rather neglected idea until Friedman's work brought it back to life and under discussion. We found similar alternative views of the a priori in Cassirer, Lewis, Pap, Carnap, and Kuhn, and after comparing the different versions, I argue for a pragmatic conception that emphasizes the constitutive elements in science without reference to an a priori because the term 'a priori' is misleading given that what is called the dynamic or pragmatic or functional a priori is not actually a priori at all in the traditional sense. The real point is that we have various theories of the constitutive elements in science, Kant's, in which the constitutive elements actually are a priori, that is, necessary and fixed, and the others, in which at least some of the constitutive elements are not fixed, so that we can understand conceptual change in science as changes in these constitutive elements. Divorcing the issue of conceptual change from a priori knowledge brings considerable clarity and moves the discussion beyond the issues of empiricism and rationalism and toward an analysis of science as a practice. While still emphasizing the importance of the constitutive element in science, I stay closer to naturalism than to the neo-Kantian position advocated by Friedman, holding a pragmatic view that the constitutive elements are principles and theories that are necessary preconditions for the possibility of a science.

A pragmatic understanding of the constitutive elements in science should be seen as a deflationary strategy that leaves a core of beliefs about constitutive elements in place, but refuses to see them as supernatural or otherwise outside of the realm of science. Regarding constitutive elements of science, we can agree on the need to begin our inquiry from some principles, but we do not need to claim that they are certain, or that they are known by some special intuition, or even that they inhabit a philosophical arena that is separated from science. Constitutive elements can function as what used to be called a priori knowledge even if they are no longer a priori in the traditional sense. On the other hand, we cannot claim that these constitutive elements are just like any other part of empirical inquiry, given that they make science possible, playing a special role because they are an uneliminable part of an empirical, physical theory, even if they are not directly testable.

My own starting point, my fundamental presupposition, as it were, is fallibilism. If we are consistently fallibilist, we will see that the search for universal principles—some fundamental axioms that are true for all time and for all places that can serve as a foundation—is a completely misguided project. Furthermore, we do not need absolutes anyway, because fallibilism does not imply that any belief is as justified as any other. There is a tremendous difference between saying that all knowledge is fallible (that we do not know anything for certain) and saying that we have no good evidence for our claims. We can rate our claims as justified or not, we can rank them in terms of margins of error that can be quantified with great precision, and we can give evidence for our claims. For example, the evidence gathered from a large double-blind study of the effectiveness of a medication is markedly stronger than a couple random testimonials that some treatment "worked for me." Of course, as fallibilists emphasize, even the best evidence can turn out to be mistaken. Nevertheless, some evidence is better than other evidence, and we rely on these distinctions in making judgments. These are commonplaces and hardly bear repeating, except that philosophers tend to forget them when they discuss relativism.

One misleading way of discussing relativism is the tendency to be concerned that relativism will leave us with no recourse in a dispute with a recalcitrant skeptic. However, no combination of data, methodological criteria, and aims would compel acceptance of a theory. No one, not even a foundationalist, is in the position of being able to rationally compel agreement. Therefore, this problem cannot discredit non-foundational philosophy, given that the same problem arises even for foundationalists. It is not as if the skeptics would change their minds, saying, "Oh, I did not realize that your beliefs are universal and absolutely certain." When one comes to any disagreement, all that anyone can do is try to give reasons for their position, based on the best evidence available and on the best standards of evidence in a given field. The foundationalists claim that their pronouncements are certain, yet that will not convince the skeptic any more than the arguments of a fallibilist, which shows that the traditional methodologist would be in no better position to defend science than the fallibilist, even if a priorism were not discredited.

As argued in Chapter 7, there can be independent constraints on theory choice even if we accept the view that all observation is theory laden. These constraints can result in objective choice because the theories that are presupposed can be epistemically independent of those under test. Furthermore, the fact that our knowledge is situated in a particular time and place, that we use a particular vocabulary, and that we begin with a set of presuppositions does not imply that our knowledge is relative. The final outcome of a scientific practice includes the results of interaction with the material world; it is not made up whole cloth by scientists. I am not claiming that all scientific practice will converge on a single answer, as a realist might; rather, I am merely pointing out that scientists are constrained by what they are actually able to construct materially. Nevertheless, we cannot escape the fact that our knowledge is the product of individuals working in a particular context, at a particular time and place, and with the conceptual and material tools that they have inherited from their predecessors. When following historians of science in attempting to understand the views of scientists from other periods on their own terms, we do not need to become relativists, even moderate ones, to take historical views seriously. If we are fallibilists, we accept that there are no universal and fixed foundations for our knowledge, and we can therefore understand how there could be conceptual change in science when the foundations are changed; however, we do need to acknowledge the special role that constitutive elements in science play in order to understand the conceptual changes that can occur during a scientific revolution.

BIBLIOGRAPHY

Nickles, Thomas. 2014. "Scientific Revolutions." *The Stanford Encyclopedia of Philosophy* (Summer 2014 Edition), edited by Edward N. Zalta. http://plato.stanford.edu/archives/sum2014/entries/scientific-revolutions/.

Peirce, Charles Sanders. 1950. "The Scientific Attitude and Fallibilism." In *Philosophical Writings of Peirce*, edited by Justus Buchler. New York: Dover, 42–59.

Index

Page numbers followed by *f* indicate a figure on the corresponding page.